598.43 Wild
Wild, Dennis

The double-crested cormorant :
symbol of ecological conflict

APR -- 2012

The Double-Crested Cormorant

The Double-Crested Cormorant

Symbol of Ecological Conflict

Dennis Wild

The University of Michigan Press • *Ann Arbor*

Copyright © by the University of Michigan 2012
All rights reserved

This book may not be reproduced, in whole or in part, including illustrations, in any form (beyond that copying permitted by Sections 107 and 108 of the U.S. Copyright Law and except by reviewers for the public press), without written permission from the publisher.

Published in the United States of America by
The University of Michigan Press
Manufactured in the United States of America
♾ Printed on acid-free paper

2015   2014   2013   2012   4 3 2 1

A CIP catalog record for this book is available from the British Library.

Library of Congress Cataloging-in-Publication Data

Wild, Dennis.
   The double-crested cormorant : symbol of ecological conflict / Dennis Wild.
      p. cm.
   Includes bibliographical references and index.
   ISBN 978-0-472-11763-5 (cloth : alk. paper) — ISBN 978-0-472-02812-2 (e-book)
     1. Double-crested cormorant—North America. 2. Double-crested cormorant—Economic aspects—North America. 3. Bird pests—North America. 4. Human-animal relationships—North America. I. Title.
   QL696.P4745W55   2012
   598.4'3—dc23   2011043626

Text design by Jillian Downey
Typesetting by Delmastype, Ann Arbor, Michigan
Font: Adobe Garamond

Claude Garamond's sixteenth-century types were modeled on those of Venetian printers from the end of the previous century. Adobe designer Robert Slimbach based his Adobe Garamond roman typefaces on the original Garamond types, and based his italics on types by Robert Granjon, a contemporary of Garamond's. Slimbach's Adobe Garamond was released in 1989.

                                              —courtesy www.adobe.com

TO MY WIFE, ELLEN,
for all her patience, encouragement, continuous support,
and belief in the book from the very beginning

TO BILL HOUGH AND CATHERINE CRAMER,
former publisher and editor of *On The Water* magazine,
for "getting" and publishing what I wrote and knowing
when there was more to write

TO SKIP PRESS,
Hollywood script guru and author, who took the time to
help me understand the value of good storytelling

Contents

Preface ix   *Acknowledgments* xv

1. Assault on Little Galloo Island: An Act of Desperation   1

    PART 1. *Legend and Conqueror of Sky, Sea, and Land*
2. From Mysterious Rescuer to a Partner to a Fowl Curse   21
3. Flight: When a Dinosaur Looked Down upon Gravity   28
4. A Time to Sink *and* Swim   41
5. The Face of Extinction: Eggshells versus DDT   52

    PART 2. *The Great Lakes: A Place of Conflict*
6. Champlain, Native Peoples, and Henderson Harbor   71
7. Fishing America's "Fifth Coast"   80
8. Ashmole's Halo: A Righteous Model of What Should Have Happened   93
9. Little Galloo Island Revisited: Praise and Outrage   103

    PART 3. *Cormorants and the Law*
10. Treaties and Legislation: War of the Wilderness   127
11. Agencies and Wildlife Conservation at Work: Cormorants in a Vise   136

PART 4. *The Channel Cat Comes of Age*
12. Catfish on a Shoestring: A Primer   155
13. The Annual Battle of the Ponds   173

PART 5. *The Past as a Clue to the Future*
14. Fortune, Timelines, and Their Intersections: When Worlds Collide   191
15. Concessions and Conclusions   198

Notes   219   *Bibliography*   229
Index   241   *Illustrations following page*   102

Preface

The word *cormorant* isn't often the first thing on people's minds when they wake in the morning and put on the coffee, but that is exactly how I began writing this book. The process itself really started almost a year earlier as an article in *On the Water* (*OTW*), an outdoor magazine published on Cape Cod, Massachusetts, featuring fishing tales, techniques, and locations throughout New England. My writing career opened with *OTW* when I submitted a five-hundred-word piece for their reader contribution column tucked away on the last page, after the classifieds and list of advertisers. The magazine had only been on the stand for a few months when I discovered it at a little market on the Cape while getting an early breakfast before hitting the morning surf in search of hungry bluefish. I guess a lot of things start with coffee in the morning.

As it turned out, my little article about fishing for bluefish from an inflatable boat fit in well with a feature story written by one of *OTW*'s regulars. The editor added a few file photos, and my little piece wound up at the magazine's centerfold with the other story rather than hidden on the last page. To convince myself that getting published was no fluke, I submitted another piece to *OTW,* titled "Weird Days," about the odd mix of fellow fishermen and onlookers I met while surfcasting Cape Cod beaches. It told of the couple from Kansas who would not believe that creatures like toothy bluefish really existed; the woman from Boston in silver high heels who asked "Who's going to clean up this mess?" when she saw the thousands of stranded baitfish that marauding bluefish had driven onto the beach; and then the gentleman, decked out in dazzling cruise wear, weighed down with bulky gold chains around his neck, oozing pungent cologne, and puffing on a massive cigar, who explained to me how to land the twelve-pound bluefish I was fighting at the time. *On the Water* published it and things went on from there.

Over time, I wrote more than thirty assigned and spec articles for the editors of *OTW,* who now and then pushed me to expand an article, making *it* a better read and *me* a better writer. My byline pieces included some how-to and destination articles along with quite a few "Zen of fishing" pieces like "Quiet Places," "Why Do We Fish?" and "Fishing Superstitions," an *OTW* online fishing tips column, and even a feature article on plankton using student drawings from a Cape Cod charter school website. My articles also appeared in magazines such as *American Angler* and *New England Game and Fish.* I later produced work-for-hire articles for a national buying consortium representing about two hundred sporting goods retailers.

Of my articles I will always consider my *OTW* cormorant piece in a class by itself. My own first cormorant experience came while my wife and I were ghosting along in light winds on Cape Cod's Little Pleasant Bay in our fourteen-foot sailboat. A startled double-crested cormorant surfacing from a dive rocketed directly into the air off our bow like a submarine-launched missile. When the color returned to our faces we laughed. The "incident" stayed with me when I plotted the *OTW* article.

I wrote the cormorant piece as a response to what I thought was a totally barbaric, inhumane, shotgun slaughter of close to a thousand adult cormorants and their defenseless chicks by greedy Great Lakes fishermen almost ten years earlier, in 1998. The shooters eventually claimed the birds were destroying their charter boat fishing industry by every year devouring tons of their clients' favorite game fish, the smallmouth bass.

In my initial research I learned that tens of thousands of breeding cormorants had set up seasonal housekeeping on New York State's Little Galloo (pronounced Ga-lew) Island in the eastern basin of Lake Ontario. In my mind, I wondered how much damage these birds could do situated on an isolated rock in the middle of one of America's largest inland freshwater seas. But feelings directed toward cormorants aren't always generated by what they actually do. Cormorants are often billed as "the most hated birds in the world" in newspapers and magazines. And considering the evil reputations assigned to carrion-eating vultures and crows, and urban pigeons, the "rats with wings," being rated as the "most hated" is not an easy accomplishment. Cormorants are despised in marinas, reviled by commercial and sport fishermen, detested by aquaculture operators, and often found insufferable by members of the public as a result of poor publicity management by the birds.

The hate also stems from people's unfamiliarity with cormorants. Diane Pence, a biologist with the US Fish and Wildlife Service, is quoted in an Internet interview: "Cormorants went away for a generation of people and now they're back. And so we have a generation that hasn't experienced the number of cormorants that used to exist."[1]

To the local fishermen, politicians, and business leaders along the shores of Lake Ontario, cormorants were plundering a resource the fishermen claimed in total as their own, their very livelihood in fact, so the cormorants nesting on Little Galloo had to go, one way or another. As an outsider, I wrote an article defending the birds and generally portrayed the charter skippers and marina owners as greedy anticonservationist killers of wildlife who did not, or would not, distinguish between their perception of the problem and the reality of the birds' right to feed in the wild. In my mind, and in the minds of media reporters, the charter boat skippers never considered the possible effects of their own overfishing, the pollution from development of the lakeshores and riverbanks, and the growing lack of biological fertility of the lake. For the commercial and sportfishermen, it was far easier to blame the ever-present cormorant for their problems instead of really looking at the effects their industry was having on the ecology and fisheries of Lake Ontario and the other Great Lakes.

By the time I finished the article, my research folder for the piece was three inches thick with reprints of articles, reports, and symposia related to the double-crested cormorant, its natural history, and, of course, its economic impact. And at that point, I hadn't even begun to look at the squabble double-crested cormorants had gotten themselves into in the South, where American catfish producers confronted hungry cormorants raiding their shallow catfish ponds every winter.

Shortly after the cormorant article appeared in *OTW*, I began to think about the many other complex issues and conflicts surrounding cormorants and how these birds fit into the attitudes and writings of newspaper reporters, prosecutors, sportfishermen, catfish farmers, conservationists, birders, state and federal legislators, regulators, law enforcement officers, and the general public. Each of these constituent groups, stakeholders, had their own view of how the double-crested cormorant was incorporated into or precipitated the conflict, producing "solutions" as varied as the number of groups involved. And in addition to individual states reacting to cormorant "infestations," I realized that cormorant conflicts extended beyond US borders, particularly involving our northern neighbor, Canada, with which we share thousands of miles of Great Lakes shoreline and which also faces many of the same cormorant issues as does the United States.

My single research folder soon expanded to several folders, then many folders, a file box, a few file boxes. . . . You can see where this was going. I saw it as a book.

With a greater realization of the number of broad cormorant issues, and maybe not with an initial full understanding of each one, I set out to write a balanced account of the cormorant using a few self-imposed guidelines. I was determined not to fall into the trap of portraying cormorants either as warm, cuddly pets or as dirty, fish-stealing culprits determined to destroy human financial interests and perhaps even civilization as we know it. On the other side, I did set out to understand the goals and attitudes of the fishermen, catfish growers, conservationists, wildlife biologists and managers, politicians, and others, without characterizing their moral fiber or assigning *unexplained* malicious intentions. I emphasized the term *unexplained,* since I was determined to write a balanced account, representing different views, not necessarily letting them pass without criticism. As an outsider I will never have a direct stake, financial or otherwise, in the many cormorant conflicts, but I wanted to understand the people who did. People and cormorants did what they did—and were what they were—for a reason. We'll explore those reasons throughout the book.

The Double-Crested Cormorant uses the illegal shooting at Little Galloo Island as a focal point and introduction to the cormorant and its controversies. From there the book looks at the bird itself, discussing its complex evolution and its many physical and physiological adaptations to flight and swimming, the bird's diving and fishing prowess, and its powerful nesting drive and unwavering passion for parenthood that made it such a successful and prolific species. The double-crested is a migratory bird, protected now by state and federal statutes as well as by international treaties. To understand these issues the book examines the roles of various government agencies such as the Department of the Interior, the Department of Agriculture, their branches and domains, and, of course, the conservation groups opposing many of their actions.

The Double-Crested Cormorant also discusses the deadly implications of the massive application of the persistent insecticide DDT beginning in the 1940s and how the chemical decimated and nearly exterminated entire populations of fish-eating birds. Then, at the southern end of the cormorant's migration, the book scrutinizes how the bird found itself entwined in the growing catfish-farming industry, facing gunners protecting their investments and livelihoods. Finally, I show how the timelines of cormorants and human interests repeatedly intersect at conflict points and look at where they are possibly headed in the future and what controversies are left brewing out there.

Sources of information on cormorants and the troubles they get into are not difficult to find. By their nature these birds either impress or aggravate a great many people. The Internet offers access to archived magazine and newspaper articles from the *New York Times* and other publications. I used public, community college, and university library sites to retrieve original texts and abstracts of studies published in peer-reviewed scientific journals, otherwise only accessible through subscription or membership in scientific societies. Where original sources were unavailable I chose what I considered reputable secondary sources that evaluated and summarized reports and studies. In some situations, I contacted and interviewed experts or participants in programs by phone to attach a human, rather than an institutional, perspective to the cormorant conflict. Also, several books listed in the bibliography provided the historical and regional backgrounds and settings for some of the difficulties into which cormorants have got themselves snarled.

Quoting Linda Wires, a research associate whose work we'll examine later, "Fiction becomes fact, and after a while even biologists start buying it. The cormorant lives in an Orwellian universe."[2] The overall intention of the book, then, is to understand the cormorant's universe, explore the relationships between cormorants and humans, and then try to clarify the conflicts for myself and my readers.

So, this June morning, while the incoming tide floods the salt marsh bordering my Cape Cod vacation rental, and as the marsh's sandpipers, plovers, and killdeer stir in the grass and a few cormorants fish the tidal creek, with a fresh coffee refill in hand, I will begin sorting it all out.

Acknowledgments

The text of a book like *The Double-Crested Cormorant* does not stand alone without the informational "spine" provided by professionals in their fields. This includes spokespeople and biologists from conservation groups, state and federal agencies, and individuals who shared their specialized expertise and directed me to others for still more details. Not to diminish the information provided by others, I especially want to thank Jerry Feist, Wildlife Biologist at USDA/APHIS/WS, who went out of his way to help me understand what cormorants meant to catfish producers; Mitchell Franz, Great Lakes charter captain and one of the activist shooters on Little Galloo Island, whose story put at least some of the human-cormorant conflict in perspective; and Stephanie Boyles, Wildlife Scientist for the Humane Society of the United States, for insights into the perceived and genuine causes of declines in sport fish populations and for reintroducing me to the idea of Aldo Leopold's "land ethic." Last, I want to thank Linda Wires, Research Associate at the University of Minnesota, for her information on the historic distribution of double-crested cormorants.

I

Assault on Little Galloo Island

An Act of Desperation

IN THE LONG RUN, 1998 was probably an average year, but it did have its own "firsts," its own records set, and its own claims to fame. In the world at large, Serbs and ethnic Albanians fought bloody battles in Kosovo; neighboring adversaries, India and Pakistan, detonated several nuclear devices in multiple tests; and three hundred million Europeans living in eleven countries agreed to deal in a single currency, the euro. In the United States, the House of Representatives impeached President Bill Clinton for lying about his affair with White House intern Monica Lewinsky and the New York Yankees claimed the World Series championship by defeating the San Diego Padres four games to zero. In theaters, the movie *Titanic* became the highest-grossing film ever, winning eleven Academy Awards. And on the small screen, seventy-six million American TV viewers watched the final episode of *Seinfeld*. Looking at the US economy in 1998, we paid just 32 cents for first-class stamps and $1.03 for a gallon of regular gas, while only 4.7 percent of us showed up in government unemployment statistics, less than half of what we counted in 2010. On a far, far smaller scale, the sport fishing boat trips and harvests on Lake Ontario plummeted for still another season. And a person or persons unknown illegally killed nearly a thousand federally protected double-crested cormorants on Little Galloo Island.

Besides the growing scarcity of game fish and fishermen, everything seemed the same as the previous year in New York's upstate village of Henderson Harbor on Lake Ontario. The same was true the year before that as well, with fewer recreational fishermen coming to town and apparently lighter weight fishing coolers being unloaded at the dock. Business owners looked at their "Beat Yesterday" sales ledgers and saw those black-ink figures shrinking. More and more often, daily and weekly figures were entered in red, indicating decreases, and owners soon reconsidered their summer payroll budgets. So not everything was the same. The fish were disappearing, and there were more of those "damn fish-eating cormorants" on Little Galloo than anyone in town could remember. The idea that maybe somebody should do something about that circulated throughout Henderson Harbor.

Not everyone within a modest radius of Little Galloo was happy about recent conditions on and about the island. Contentment was probably as far from what was in the minds of local residents as one could imagine. The fact that thousands of birds had taken a shine to summering on Little Galloo was not really the problem. Generally, most people like birds. But not these birds. These birds, double-crested cormorants, ate fish and lots of them. But birds that eat fish are not always considered a problem. Seagulls, for instance, feed regularly on fish and in some coastal areas they are not well-liked, but they are tolerated, because again, in general, most people like birds. The difficulty that arises with cormorants is that they often feed on the same fish as humans do, the same fish, in fact, that some humans, fishermen, pay other humans, charter boat captains, money to help catch them. Cormorants soon became the enemy, and enemies meant conflicts.

Charter boat fishing in eastern Lake Ontario wasn't just a business. It was *the* business. Henderson Harbor was a fishing village located just a few miles from Little Galloo's nesting cormorants. The village at one time proclaimed itself the "Home of the Black Bass" and had erected signs at entrances to the town certifying it.[1] A species of black bass, the smallmouth bass, was one of the top fish sought by Lake Ontario anglers. Henderson Harbor continued to exist as a village and a political entity primarily due to the smallmouth bass and the sportfishing industry associated with it. Without the influx of fishing dollars spent on bait, tackle, and charters, and in the restaurants, taverns, marinas, and motels, Henderson Harbor would not be what it was: a fishing destination. Without the fish there would be no fishing, no fishermen, and no fishing dollars. Business owners would then have to cut their seasonal payrolls, thereby reducing the town's total disposable income for the season and the year.

It was said by the charter captains that the growing numbers of fish-eating cormorants on Little Galloo Island hunted and preyed on sizes of smallmouths up to twelve inches in length, the same length as New York State's legal minimum size limit for anglers. Humans and cormorants were fishing for the same fish, driving many fishermen to feel that, left unchecked, double-crested cormorants could severely damage their industry and their income.

Mitchell Franz was one of Henderson Harbor's many charter boat captains at the time. Working in a group with other captains and business owners, Franz repeatedly contacted officials of the US Fish and Wildlife Service (USFWS or FWS), requesting help in the form of control measures for the growing numbers of cormorants. Franz reckoned that in 1980 the Little Galloo Island breeding cormorant population was about 200; by 1998 there were 9,800 pairs. The birds were doubling their numbers every three years. Casual observations of the birds told the story of their swelling population. Franz recounted how local cormorants were so numerous that the air was thick with them and how continuous flocks "flew past for an hour or longer to feed in the bay."[2]

Captain Franz made the observation that between the late 1980s and the early 1990s, Lake Ontario was an "excellent fishery for smallmouth bass, steelhead, and brown trout." He believed that as the cormorant population grew the fishing declined and that the two changes were related. He told of how trout fingerlings were once "shore stocked" from tanker trucks and by the next day hundreds if not thousands of cormorants would show up and finish off the entire stocking. Franz added that the birds were eating over a million smallmouths a year and millions of yellow perch.[3]

As a result of the declining sportfishing, Franz and other interested parties found themselves "going to a lot of meetings" but with nothing happening. Captain Franz felt that federal agency representatives showed up at meetings unprepared, different representatives each time, so there was little continuity and as a result a lot of time was wasted with "the feds." The local citizens group presented a request that two recommendations be implemented: (1) fund a study of cormorant diets to determine what fish they actually eat; and (2) make a determination of how many cormorants Little Galloo Island could support. Franz recalled that no action was taken on the recommendations. (Actually, New York State Department of Environmental Conservation biologists at the time were studying regurgitated double-crested cormorant "pellets" to learn what the birds ate.) Franz also said he was shocked when one USFWS official stated at a meeting that if the birds were eating as many fish as claimed by the fishermen, the USFWS needed to do something about the angler take, interpreted by the anglers to mean that the birds had more of a right to fish the lake than the fishermen did. In a published retelling of the encounter after the shooting, Franz remarked, "That's what loaded the gun."[4] Again, in Franz's account, another response from "the feds" was that double-crested cormorants, as migratory birds, were specifically protected from lethal controls by international treaty. The captains were subsequently told by agency representatives that nothing could be done.

And of course there is the other side of the tale of these meetings. Diane Pence of the USFWS attended many of the meetings Franz refers to leading up to the Little Galloo incident, but her recollection is at odds with Franz's version. In a phone conservation, when asked if the captains were being listened to she responded, "That's a good question, but they were being listened to, but they didn't like what I had to say." Pence remarked that many times the general public and the recreational anglers misunderstood cormorants and were reluctant to change their opinions even in the face of evidence presented to them. Pence felt the meetings were not balanced in that they were attended by "one likeminded constituency. They were all anglers and charter boat captains." Missing were neutral parties such as independent conservation group representatives speaking to achieve a balance of opinions.[5] Many meetings were attended by then congressman John McHugh, whose district included the town of Henderson and the village of Henderson Harbor. As a strong advocate of the fishermen, the congressman was known for powerful, anticormorant rhetoric, claiming that *each* four-pound cormorant ate several pounds of valuable fish every day, as opposed to the single pound they actually ate. It was said that McHugh and his staff likened the Little Galloo cormorant incident to a shot across the bow and a message to the Fish and Wildlife Service.

Both parties finished the series of meetings feeling that little had been accomplished. Peace was at risk. War was on the horizon.

As one fishing guide put it, "We've got no Kodak, no DuPont. You're either a fisherman or a farmer if you're going to live here."[6] And the captains were no farmers. They were dyed-in-the-wool fishermen, resolved to protect their own interests against the threat, real or perceived, of cormorants ruining their livelihood.

The deadly challenge to the cormorants' secure nesting on Little Galloo came in late July of 1998. In the dark of the night five men set out from a marina in Henderson Harbor. As they boarded the boat they passed shotguns and ammunition over the gunwale and stowed them for the short run to Little Galloo Island. After what they felt was a long period of neglect and animosity by federal agencies, the five men had earlier decided to take action on their own. The captains believed the federal agencies that were supposed to protect their interests had failed them in favor of protecting the birds.

The boat soon nosed up to the darkened shore of Little Galloo Island. Three of the five men stepped ashore carrying twelve-gauge shotguns and sufficient shells to get the job done. The northern area of upstate New York is known for its duck and goose hunting in the fall, so guns, ammunition, and bird shooting were nothing new to these men. The two men remaining aboard backed the boat off the shore and circled the island. The island's nesting cormorants were still unaware of the terror to come that night.

On that same July day the seven or eight thousand pairs of breeding double-crested cormorants nesting on Little Galloo Island cared little for the price of gasoline, Oscars, or the President's affairs. They had settled on Little Galloo to raise their families.

Cormorants were not strangers to the Great Lakes. Regardless of the folklore tales and angler views, double-crested cormorants *are* native to America. For thousands of years they spent the warmer months on their breeding grounds on the rivers and lakes of the Midwest. And then in the early 1900s cormorants showed up on the more isolated islands of the Great Lakes basin, where they had probably also nested hundreds of years ago. In the media, among anglers, and among other interests focused on fish and fishing, cormorants were somehow considered and labeled foreign, exotic, and nonnative or even thought of as an invasive species. It's far easier to demonize an alien species and garner support in battling an invasive species perceived as ruining such a traditional enterprise as fishing than it is to produce sympathy for anglers competing with a truly American bird. To a large degree, the conflict is about publicity, and in fishing communities cormorants got nothing but bad press.

Archaeological and other records document pre-1900 nesting grounds along the shores of the Great Lakes and areas of the Midwest, in which cormorants were far more abundant than they are even today. Huge flocks of migrating double-crested cormorants were described in terms of square miles of sky. As early as 1926, flocks estimated at hundreds of thousands of individual birds moved north along the Mississippi River. With these numbers in mind it's difficult to accept the rationalization offered later by fishermen and others that an "overabundance" of cormorants today was causing an imbalance in the ecology when, in fact, modern double-crested populations represented actual *declines* from historic highs.

Use of the term *overabundant* is a value-laden judgment statement; it implies that cormorant numbers exceeded the biological carrying capacity of their food supplies. But, as we'll see in the upcoming discussion of Ashmole's Halo, it is the food supply at breeding sites that acts as a regulator of populations, allowing them to grow only as large as the local food supply can support. It can then be argued that these millions of birds thrived due to an existing ecological balance between their numbers and the fish on which they fed; otherwise there would be no possible expansion of the cormorant's population. It's therefore possible, in an absolute scenario, that humans, as fishermen, may be the intruder, the "invasive species" disrupting the stability of a balanced ecosystem, not the double-crested cormorant.

Regardless of how or when cormorants settled in the area, the massive feeding potential of the Great Lakes allowed the species to grow to its present-day levels. To many residents the huge and growing flocks of cormorants were still unwelcome foreign visitors. And through several peaks and plunges in their population, successful pairs of breeding cormorants came to colonize and virtually commandeer numerous Great Lakes islands. One such island was Little Galloo.

Sitting in the eastern basin of Lake Ontario, a few miles offshore from the fishing town of Henderson Harbor, Little Galloo, an unremarkable half-mile-long rock pile, was one of the most productive double-crested cormorant rookeries in the United States. In the summer of 1998, and in recent summers past, the island rookery was jammed with nests: gull and tern nests in the sand and the flat, bulky nests of double-crested cormorants almost everywhere else. On the ground, among the weather-worn rocks, boulders, and few remaining grasses, cormorant nests pervaded Little Galloo's landscape. Even overhead, in tree limbs stripped bare of their leaves and twigs by the birds, cormorants wedged their nests into every available support strong enough to bear the weight of the nest, the two four-pound parents, and their chicks. Double-crested cormorants were also not opposed to building their nests atop the huge guano mounds that had accumulated for years beneath the dead nesting trees. In the birds' world they had come to reproduce the species' genes, not to fret about the housing situation on Little Galloo Island.

Cormorants had claimed much of Little Galloo as their own, but the island rookery was just part of the bird's story.

For a birder or naturalist, double-crested cormorants are not difficult to find. Often misidentified from a distance as ducks, loons, or even geese, they are seen flying at practical mid-heights or more commonly skimming the water's surface in straight-line formations, ragged-V wedges, or in large curved arcs. Unlike the identifiable quacks and honks of ducks and geese announcing their approach, cormorants, whose voice is a series of deep guttural grunts, choose a stealthier advance and remain silent when they fly. Another sighting might find cormorants paddling across the surface of the water, followed by the flurry of snappy dives in pursuit of prey. Or binocular optics might scan a cormorant popping to the surface while swallowing its new struggling catch head first.

Above the water, cormorants travel through the air, but compared to many other birds, the cormorant is considered only a moderately efficient flyer, meaning that flight consumes a great deal of the bird's hard-earned energy. Because of this, cormorants are apt to perch, roost, and nest close to their most productive feeding areas. So after hunting, it is not uncommon to spy clusters of cormorants perched on nearby pilings, buoys, boats, docks, or rocks, faced into the wind, with their wings extended in a spread-eagle position, drying them in a fair breeze.

It's easy to see how, in the course of only an hour or so, cormorants might visit and face the challenges of each of their four environmental worlds: the air, the land, the water's surface, and its depths.

In appearance, the double-crested cormorant, sometimes referred to by naturalists and researchers as the "DCCO" for shorthand purposes, is a large black bird with an overall greenish gloss. Its scientific name, its genus and species, is *Phalacrocorax auritus*. The genus name *Phalacrocorax* is interpreted to mean "bald-headed raven" from the Greek, and the species name, *auritus*, is from the Latin term for "eared," referring to the short tuft of feathers that develops above the eyes of adults of both sexes during the breeding season.[7] Including its snakelike neck, adults stand twenty-nine to thirty-six inches tall, sport wingspans of up to fifty-two inches, and weigh about four pounds. The bird's terminally hooked, mottled orange bill, sometimes mentioned in folklore as presenting a "bemused grin," is surrounded at its base by an orange or yellowish patch of naked skin called the gular (pronounced goo-lurr) pouch. The double-crested's only other distinct coloration appears in its small, bright, alert, turquoise-green eyes, which, contrasted against its dark plumage, no doubt adds to its characterization as a gothic figure. When either swimming on the surface or diving to surprising depths, cormorants propel themselves through the water, wings at their sides, by means of their characteristically webbed feet.

The double-crested cormorant at first glance fits what we think of as the standard model of a bird. It has feathers and wings, feet, a head, a tail, and lays eggs. But because the cormorant has adapted to several environments, it became a very complex organism. If we first understand its natural history, behavior, and a touch of its unique anatomy and physiology, we'll pave the way for appreciating how cormorants live and seemingly compete so well with humans.

As one would guess from where they live and feed, and how they make a living, cormorants are excellent swimmers and divers. They can remain submerged for as long as seventy seconds, and perhaps longer, with dive times averaging twenty to thirty seconds, depending on the depth, speed, and density of their prey. As for "holding their breath," especially since they spend so much of their time under water, cormorant lungs, compared to those of many terrestrial animals, are small in proportion to their body size. In addition to lungs, the cormorant's respiratory system, similar to other birds' but modified for diving, also includes a unique network of interconnected functional air sacs, expanding the cormorant's oxygen-carrying capacity.

Cormorants are classified as waterbirds rather than seabirds. They are not pelagic or sea-roving birds. They can dive easily to the twenty-foot mark and may reach as far down as seventy-five feet when necessary, but because of these limitations cormorants are unable to hunt in the vast depths of the open sea, where food is scarce. Therefore, they congregate along shorelines, islands, and estuaries, in shallower water, usually with sloping bottoms supporting plentiful food supplies. Also, because of their need to dry their wings, these birds require a nearby off-the-water structure on which to perch and expose their wet feathers to drying breezes.

In between wing-drying and perching periods cormorants hunt for fish. The cormorant's fishing process has been witnessed and recorded by many investigators and naturalists. During hunts, they target small schooling fish beneath the surface, as a rule fish four to six inches long. The birds seize their prey in their hooked bills and may even swallow smaller fish under water. They swallow larger or more difficult spiny fish on the surface, where they sometimes flip the prey to align it properly for a head-first swallow. On occasion, cormorants have been observed taking larger prey up to a foot in length, but without exercising caution it's possible for the bird to choke to death trying to swallow it. As fishers, double-crested cormorants are opportunistic feeders, using trial-and-error searches and often foraging where they previously found prey. Since the hunt is essentially a numbers game, schooling fish present better prospects than individuals, in that schools offer more targets and better odds of success, generating fewer "empty-handed" trips back to the surface.

And how much fish does a cormorant consume? Several scientific studies show that over the course of a day a cormorant consumes about a pound of fish, approximately 25 percent of its body weight. The one-pound figure is an average number. There may be times when forage fish are scarce for one reason or another, limiting their intake and lowering the bird's consumption. At other times, when massive numbers of fish gather on their spring spawning grounds or when the birds ready themselves for an energy-consuming migration, hungry cormorants may increase their calorie intake and devour up to a pound and half of fish in a day.

The types and species of fish that find their way into cormorant bellies is a contentious issue. Much of the controversy that surrounds the cormorant stems from what it is thought to eat. We'll examine this topic in more detail in the chapters concerning the bird's conflicts with Great Lakes fishermen and southern catfish farmers, but the subject is important here as well. Fishermen and aquaculture producers claim that every year cormorants swallow hundreds of tons of commercially valuable smallmouth bass, trout, salmon, walleye, perch, and farmed catfish. In truth, cormorants are not at all choosy about what fish they pursue; they'll accept what's offered as long as the prey size is right and there are enough of them to make diving efforts and pursuits worthwhile. As long as the fish are there, the cormorants will come.

For the most part, the cormorant's diet reflects the body of water in which they feed. Since cormorants inhabit saltwater and freshwater shores, as well as brackish estuaries, their diet varies with each group's environment. On the Great Lakes, cormorants favor rainbow smelt, alewife, young smallmouth bass, various panfish species such as yellow perch and bluegills, and more recently an invasive species, the round goby. Like the goby and the alewife, a large portion of the bird's prey species, particularly in the Great Lakes, are nonnative fish. These invasive species were probably introduced by contaminated ballast water being pumped into American waterways by foreign ships entering the lakes through the St. Lawrence Seaway and via various canals. On saltwater shores, cormorants feed on menhaden (called *bunker* or pogies by anglers), black eels, sand eels, flounders, and any one of the myriad of schooling forage fish found in salt marshes and bays. And, of course, at the southern aquaculture ponds the menu consists primarily of assorted sizes of cultured channel catfish. In saltwater or fresh, cormorants get into trouble by pursuing many fingerling stages of native species, the older generations of which sportsmen heavily target, spend their fishing dollars to catch, and pursue to fill their coolers. Given all these food fish common to man and cormorant, conflict, sometimes with violent results, was, and is, inevitable.

The cormorant's fishing efficiency no doubt contributes to the human-cormorant conflict. Anglers may even be jealous of the bird's fishing skills. Underwater, the cormorant's sleek body, long neck, and webbed feet make it a very successful hunter. But not every one of the cormorant's adaptations was positive in every way. In fact, some of the cormorant's anatomical features acted in direct opposition to the efficiency of others. The evolutionary placement of the cormorant's feet toward the rear of its body, for example, added a great deal of power to its dives but presented other problems: its wobbling walk on land and the long awkward running motion it uses to gain thrust and lift in its labored takeoff. Compared to cormorants, nondiving ducks, which have their legs located closer to the center of their bodies instead of at the rear, are capable of a nearly vertical spring into flight to quickly escape predators and other threats. Cormorants, however, can sometimes require up to a hundred yards of open water to get airborne, depending on the strength and direction of the prevailing wind. They gain speed by tiptoeing along the surface of the water and flapping their wings, with their wingtips brushing the surface until they reach takeoff velocity. The alternative to their hundred-yard dash is taking off from an elevated perch position, such as a high piling, tree, bluff, or manmade structure, with the bird initially losing height but eventually regaining altitude by means of rapid flapping.

"Flew well, landed poorly," expressing the importance of returning to the earth's surface safely, surely applies to our cormorants. Being a waterbird, just like its takeoffs, the cormorant's landings usually take place on water. They use a method often referred to as a "water ski landing." Similar to a plane landing, the cormorant raises its nose, wings, and body to increase drag and reduce speed. When the tail feathers are lowered as air brakes, the bird flaps its wings forward and backward in a hovering motion to achieve final landing velocity. Its webbed feet, the "water skis," then come into play, plowing a furrow in the water and lowering the bird to the surface. As clumsy as it looks, the cormorant's landing is a complex sequence of learned movements designed for a specific environment. We'll revisit the topic of the cormorant's long takeoffs and splashy landings in later chapters during the discussion of just how these maneuvers affect the tactics catfish farmers use to prevent the bird's costly raids on their ponds.

Like many of the fish they consume, cormorants move with the seasons. In their traditional migrations the birds leave their northern and midwestern breeding grounds in October and November to head south. When the cormorants settle into their southern winter quarters they frequently roost near the area's expansive catfish farms. Banding studies have shown that birds arriving from each northern nesting area, on the Great Lakes, for example, exhibit extensive mixing with birds from other northern groups. It's interesting to note that birds migrating from east and west of the Rocky Mountains exhibit different behavior, with little or no mixing on their wintering grounds. In late March, April, and May their migration north begins from their southern wintering grounds. Observations show that birds of different ages begin their northward exodus at different times. Birds two years old and older leave first, while birds less than one year old do not depart until later, normally in May. This succession of migratory waves assures that older, breeding adults will arrive on their summer grounds, like Little Galloo Island, before immature, nonbreeding juveniles, giving the nesting ritual a seasonal head start.

In their recurring movements double-crested cormorants travel well-established pathways along both coasts, the Mississippi River and the Central Flyway as compiled by the USFWS. On a visualized map, the cormorant's northern and southern ranges can be divided by drawing an imaginary horizontal, east-west line connecting Tennessee and Maryland. It's worthy of note that only the cormorants in the northernmost range are migratory, while many southern groups stay at home, colonize, and nest in year-round areas.

During their migration double-crested cormorants move at an impressive pace. They can advance about forty-three miles a day in flocks of up to two hundred birds, with some larger formations reaching over a thousand individuals. Migrating cormorants have hundreds of miles to fly in order to reach their wintering grounds in the fall and then return to their breeding grounds in the spring. Cormorants heading north from Florida, for instance, fly their twelve hundred air miles in large flocks to settle on Lake Ontario or the other lakes in the Great Lakes basin. Many continue on deeper into Canada, adding additional mileage. Coming from farther west, Texas birds log almost half again as many miles, while cormorants wintering in North Carolina wing only about six hundred miles or so to nest on Little Galloo Island and neighboring areas.

With such a broad migration range it's easy to see how cormorants come into conflict with human interests at so many locations. These sites include both coastal and inland locales across many states. In fact, cormorants are all but omnipresent nesters. They nest in forty-three states, every Canadian province, Cuba, and the Bahamas. So wherever cormorants winter or breed they present the opportunity to trouble humans.

Having arrived at their nesting sites, double-crested cormorants quickly commence their breeding cycle. They are colonial nesting birds and frequently nest in vast numbers with other colonial nesters, including great blue herons, green herons, night-crowned herons, brown pelicans, gulls, terns, and occasionally noncolonials like ospreys. What frequently separates cormorants from other waterbirds at these sites is their superior numbers. Because of its long, possibly twenty-year, life span, the cormorant has an enormous reproductive potential. They begin breeding at the age of three and may continue to breed through the next fifteen seasons. Even if only half of them ultimately survive, a single female cormorant might produce as many as seventy eggs throughout its reproductive life, enough to give the cormorant a commanding presence on the breeding grounds.

Cormorants are colonial in their nesting behavior, and to a degree that is universal. They certainly welcome the company of other breeding cormorants and many other bird species at their nest sites, but they are shy of humans and generally cautious of mammalian and avian predators, so cormorant nesting sites are always isolated in some way from mainland activities and civilization. Locations include sandy beaches and peninsulas, small stands of deciduous trees, trees killed by older cormorant droppings, cypress trees in marshy shallows and at wetland edges, hundred-foot-high supports and catwalks of electrical transmission towers, and rocky, nonforested islands raked clean of vegetation by winter storms.

These safe, isolated nest sites near expanding urban, suburban, and even developing rural areas are becoming scarcer. A study of Virginia cormorants conducted by the College of William and Mary's Center for Conservation Biology concluded that "the availability of 'predator free' structures may ultimately serve to limit the breeding population."[8] In North America, *P. auritus* faces a number of predators with tastes for cormorant eggs and chicks. From the skies, avian marauders such as great black-backed, herring, and ring-billed gulls, kites, eagles, turkey vultures, and crows raid their nests regularly for easy meals. Then from land and water come wild foraging river otters, raccoons, skunks, foxes, and rats, plus roving domestic and feral cats and dogs. Even at island rookeries, coyotes, having wandered over thick winter ice, become resident predators and feast throughout the nesting season.

At their protected nesting sites cormorants are monogamous but only for the current reproductive season; their mates and choice of individual nest sites change from year to year.

Most bird species engage in courtship rituals to form mating pairs, and cormorants are no different. To attract the attention of an interested female the male cormorant displays his plumage from the likely nest site he has chosen by proudly waving his outsized wings. He also lifts his bill skyward, exposing his colorful gular pouch and may initiate a gular flutter to further increase his attractiveness to females. With no possibility of being characterized as a melodic songbird, the cormorant's "tune," basically a series of grunts, is a "low, continuous *ok ok ok ok ok ok* for about an hour at thirty-second intervals."[9] The male continues his display until one female, impressed by his ritual and vocalizations, settles down at his nest site and exchanges greeting displays consisting of clicking sounds, head motions, and head and neck caresses. They are then a nesting pair for the season.

While the female guards the site her prospective mate gathers nesting material such as seaweed, feathers, sticks, rope, line, discarded plastics, and parts of abandoned fishnets. The male may even present tossed-aside debris from the sea or lake bottom collected during his forage dives, including hair combs and rusted pocketknives. One story of the thoroughness of the cormorant's recycling ability tells how a Cornell University student marked the hidden nests of common terns nesting along with the cormorants with red surveyor's flags for later study. The next day the returning student found that someone had stolen virtually every marker. On investigation he found that the crafty double-crested cormorants nesting nearby had confiscated the colorful flags and promptly woven them into their nests.[10]

Dealing with whatever materials the male supplies, the female interlaces them, and with excreta cements them together into a broad, flat nest. The entire process takes about four days. During construction and afterward, the nest is never left unguarded. Like other parents cormorants are inclined to protect their home, work, and family. In a book titled *The Submarine Bird* the author notes, "The pair defends their territory by displays of intimidation, seldom by actual combat. But the displays can be aggressive, with hissing, wing-striking, and dagger thrusts of their beaks."[11]

When nest construction is finally completed, the new pair mates at the finished site.

Shortly after mating the female lays a beautiful clutch of three to five elongated bluish eggs over a period of about three days. Her eggs, at just 2 to 3 percent, are small relative to her body size and compared to those of many other bird species. Cormorants are very good, protective parents and exhibit such powerful reproductive drives that at times when parents have had their eggs destroyed by storms, man, or predators they will pick up large pebbles, place them in their nests, and attempt to incubate them as if they were actual eggs. It's also been observed that all adult cormorants within a colony do not nest or mate on any one given schedule. Many researchers studying cormorant nesting habits noticed nests at different stages of development in the same colony. They ranged from newly constructed nests to nests filled with mature eggs to nests with recent hatchlings. This type of protective behavior may have evolved as a defense against a complete seasonal reproductive failure for the colony. A single disaster, in the form of a dramatic spike or drop in temperature or a violent storm or flood, would have the ability to kill many if not all of the young-of-the-season at the same stage of development. So the adults, not programmed to lay and care for their eggs and young at a specific time, provide built-in protection for an entire generation of their offspring.

Survival of its eggs and the conquest of the world of dry land depend strongly on the cormorant's doggedness and determined parental care. To strengthen their caregiving, cormorant parents take equal turns incubating their eggs, giving each one a respite in which to rest and feed. If the nest site was selected with care, it includes a nearby perch on which the off-duty parent can relax and still observe the nest and remain alert for predators.

Since double-crested cormorants lack brood patches, a feature common to many other birds for transferring heat to their eggs, they instead use their webbed flat feet as heat radiators to keep the developing eggs at the proper temperature. The parents virtually stand on their eggs, wrapping their webbed feet around them, for twenty-five to twenty-eight days, until the eggs hatch. At hatching each chick pecks its way through and out of the blue shell with an egg tooth, which it later loses. After the chicks hatch, the female clears the nest of egg fragments by flipping them over the side. The problem then arises that over several seasons nesting cormorants have stripped away much or all of the vegetation, leaving nests built on the bare ground or in dead trees exposed to bright sun and high summer temperatures. To protect the vulnerable chicks on the barren landscape the always-protective parents alternately shade them with their own bodies.

By mid- to late July most of the viable eggs will have hatched. The chicks at that time are still unable to drink, hunt, or feed on their own. It's up to the attentive parents to take turns carrying water in their mouths for the chicks to drink and regurgitating partially digested fish for their still flightless, defenseless hatchlings to eat. In feeding the very young chicks, adults arch their necks to engulf the chick's head with their mouths and regurgitate liquefied food. Older, more aggressive chicks feed by plunging their heads into the parent's throat and grasping whole fish, which the adult has regurgitated into its neck pouch.

Once the chicks are able to feed and drink what the parents provide, their progress toward self-sufficiency comes in leaps and bounds. Time frames vary from colony to colony, but usually at two weeks old the chicks are capable of standing most of the day. A week or so later the growing chicks leave the nest for short periods of time and begin to explore their nearby surroundings. They gather together with other chicks in groups called crèches but return to their respective nests periodically to be fed by their parents. The young birds can dive by their fourth week as part of their exploration, and even though the young are still tended by parents, they attempt flight five to six weeks after hatching. At the seven-week mark the young have abandoned the nest and fish and swim with the adults but continue to depend on their parents for food. And after another three weeks, at ten weeks old, the chicks are completely independent of their parents, fishing and feeding beside the adults.

Considering the attention cormorant parents give their young, including incubation, standing guard, shading, feeding, and watering, the mortality of newborn birds seems relatively high. Statistically, on the average, fewer than two of the three to five eggs result in chicks that survive long enough to begin the trek south the following fall. It's easy to imagine how disease, weather changes, accidents, and predators can take a certain number of the eggs and young birds, but evidently a sufficient percentage of chicks live long enough to reproduce, maintaining and, as we've seen on Little Galloo Island, increasing the species' numbers.

The cormorant's north and south migrations, along with its choice of isolated nesting sites, would clearly not be possible if at a time in its history *P. auritus* had not adapted itself to flight. Also in the cormorant's evolution the species found ways to exploit the underwater world as a source of food. The evolutionary changes required to fly and to dive are each imposing, but their combination in a single species is nothing short of amazing.

So, other than threats from predators such as gulls and crows and the occasional eagle, isolated from prowling mammalian predators, the colony on Little Galloo Island was a safe haven for young cormorants. But that security was soon to be challenged.

To the men with guns in their hands that July night cormorants weren't all that amazing and the security of the island colony was soon to be a thing of the past. The birds were nothing more than hungry competitors poaching *their* fish. And on top of everything else, the federal government was adamant in their protection. Something had to give. This night would bring their cause to the forefront of the federal agenda.

Back on Little Galloo itself, in the otherwise safe cormorant colony, adult cormorants probably stirred nervously as they heard the approaching boat and then the voices and footfalls of the three armed men coming ashore. The men had visited Little Galloo Island four or five times before to shoot some cormorants. At those times they drove and herded the birds off the island into the range of armed men firing from a nearby boat. To conceal the shootings, the men dumped the bodies of any dead birds that fell back onto the island into the water. Scavenging birds, fish, and nature worked together to destroy the evidence. This time would be different: more birds would die, and the carcasses would be left in the open as a message to conservationists and regulators to finally take action.

Moments after landing, the three shooters blasted their way through clusters of cormorant nests, killing adults and chicks alike. Terrorized adults that survived the initial onslaught abandoned their nests and chicks against their ingrained protective instincts. They scurried over the rocks, flapping their wings, gaining speed for takeoff and the apparent safety of the air.

But then even more muzzle flashes from the circling boat blazed in the night. Hundreds of shotgun pellets tore through fleeing cormorants, further thinning the flocks. The dead and wounded birds dropped into the lake and back onto Little Galloo Island, where the other three shooters continued sweeping the rookery, emptying their weapons into unprotected nests and flightless chicks.

At some point in the night the five men decided their work was done. More than eight hundred cormorants, perhaps closer to a thousand birds, protected by federal law and international treaty, lay dead and or would soon perish from their wounds. Some adults were maimed so badly they would be unable to fly, dive, or feed and faced very uncertain futures. Many abandoned chicks would soon die of starvation and dehydration. And still other injured cormorants would be euthanized by researchers who later discovered the slaughter.

Back aboard the boat, before they returned to the harbor, the five men rendezvoused with two other conspirators in a second boat, to whom they handed their weapons for safekeeping from the prying eyes of investigators they knew were sure to follow.

Like many of the other "firsts" of 1998, this was the first known, and soon to be well-publicized, assault on Little Galloo Island. It was a true act of desperation. But was it necessary? And who exactly were these men and what was their real intention? Was the shooting an act of civil disobedience or a cruel environmental crime? Was it an action based on reality or only on the perpetrators' perception of reality? These are only some of the issues and questions we'll explore in the following chapters.

PART I

Legend & Conqueror of Sky, Sea, & Land

2

From Mysterious Rescuer to a Partner to a Fowl Curse

SOCIETIES AND CIVILIZATIONS often have more shared characteristics than they do traits that isolate them. Human cultures far apart in both time and geography have a way of applying different materials to accomplish the same goal or to suit the same need. Breads, tortillas, and noodles fit the same niche in human diets depending on whether the culture and climate favor wheat, corn, or rice. Another use of grains and other staples, a thread running through most cultures, is fermentation. Various grains, fruits, and distillations produce potions such as vodka, rye whiskey, scotch, rum, mead, gin, wine, and other libations, probably as numerous as there are cultures. The point? People and their cultures use what's familiar to them, what's at hand, what's visible. This premise applies to their food, drink, music, language, literature, and folklore.

One of the variables, and, oddly enough, a consistency in human societies, is the use of birds in religion and folklore. Quetzalcoatl, the feathered serpent of the Aztec people, represented many things in their civilization. He was the god of creativity, knowledge, and fertility and the inventor of books and the calendar. He was or was related to the gods of the wind and the dawn and gave mankind the gift of maize. When Cortez entered the Aztec cities in his startling armor, Montezuma considered him to be Quetzalcoatl incarnated.

Quetzalcoatl, as a giving god, was not typical of the group. The

Mayans, in their later civilization, worshipped the fearsome bird, Camulatz, who cut the heads off the first men and devoured them. Of two bird figures of the Greeks, the Phoenix, a mythical sacred fire spirit, rose from its own ashes to live again a life as long as its previous one. The second of the Greek birdlike figures, the Harpies, perhaps the representation of storm winds or death spirits, were first portrayed as beautiful winged maidens but were later conceived of as ugly, winged, bird women with talons, who snatched away food and carried off people to the underworld. In Native American cultures it was taught that Raven, a dark, notorious shape-shifter and trickster, was, oddly enough, the creator of human beings and the firmament in which they dwelled. The trickster Raven was also the keeper of secrets, mankind's guilt, and humans' innermost, secret thoughts.

In the myths, birds or winged gods sometimes befriended human beings. In one famous legend, birds, cormorants in fact, came to the rescue of a sailor in trouble. "The Fisherman and the Cormorants of Udrost," a Norwegian folktale, tells how a poor but generous Fisherman blessed with a large family went to sea in his small fishing boat.[1] After being lost in fog and battered by a dark, angry storm, and unable to locate his home, he heard a loud shriek. Even thinking it a devil, as his only hope the Fisherman steered toward the sound and spied something black ahead. As he approached, what he saw was no devil but rather three cormorants perched on floating driftwood. Drawing even closer to the birds, his boat somehow struck land, the dark sky cleared, and the angry storm ended. It was land, but not the Fisherman's land. This new setting was the very idea of perfection; it was Udrost: the Norseman's island paradise, the perfect fairyland seen only by pious sailors in dire straits at sea.

The Fisherman was soon greeted by an Old Man with a white beard who brought the Fisherman to his home. From the outside, the hut appeared simple, but within it was a palace. The platters of food and pitchers of drink were endless. The Old Man mentioned his three sons, who the Fisherman must have passed coming to shore. The Fisherman said he had only seen three cormorants. The Old Man said that they were indeed his three sons—in search of nonbelievers. The Fisherman later fished at sea with the Old Man's three sons and filled their large boat to the rails with their catch. When it was time to return home, the Old Man told the Fisherman that to return to Udrost he was to follow the cormorants before Christmas. At that time he would be able to fish at sea and fill his own boat with fish to sell at the market for a large profit. After being home for a while, the Fisherman decided to return to Udrost before Christmas. Having heard the Fisherman's tale, his Greedy Neighbor, who wished to visit Udrost also, was told to follow the flight of the cormorants but to be cautious because the birds did not care for unbelievers. The Fisherman found Udrost and became wealthy from the sale of so many fish. The Greedy Neighbor was never seen again. The fable presented the mythical cormorants as friends of righteous fishermen but adversaries of skeptical and doubting unbelievers.

It might be interesting to consider an imaginary modern scenario of present-day cormorants confronting modern captains. Would cormorants guide the captains to the fish? Would the cormorants save fishermen, believers or not, lost in angry storms? How many fish would the fishermen put aboard their boats to sell at the market? Would the men take on the guise of the Greedy Neighbor or remain generous fishermen? We may get some of these answers later in the cormorant's story.

Setting greed and belief aside, in Europe the Irish and Welsh peoples had a place in their own mythology for the cormorant. Invoked as the magical Sea Raven, the cormorant was skilled at connecting with the natural forces in both air and water. In England the cormorant is depicted as the Liver Bird (rhymes with *driver*), the symbol of Liverpool City. As a possible heraldic representation on the royal seal, two eighteen-foot copper cormorants were erected in 1911 on the towers of the city's Liver Building. The folklore of the city says that the cormorants are male and female; the female looks seaward in search of returning seamen while the male looks toward the city itself, confirming that the pubs are indeed open.

Another legend, that of Count Dracula, the vampire well known in print and film, was brought to life by the Irish writer Bram Stoker in his novel of the same name. Stoker may have conjured up the image of the transformation of the vampire into a cormorant rather than into the more familiar bat. One literary essay has analyzed the transformation of Lucy Westenra, one of Dracula's victims, as she becomes less and less human and joins the ranks of the "undead." After Lucy is first bitten by the count, she writes to her friend that she has developed an appetite like that of a cormorant, and through Stoker's words Lucy relates how she felt she was sinking into green water and later flew up from the water's surface to pass over the local lighthouse. The essay's author draws the possible parallel that cormorants "sink" into the water to feed on fish and later fly from the surface to lighthouses, where they often choose to roost. It may be that cormorants also "never drink wine" and dislike mirrors, crucifixes, and garlic, but we'll need another sequel and additional research to answer that question.

Cormorants of legend may not necessarily have a warm place in their hearts for unbelievers, but an instance of a warm heart for cormorants appeared fifty years ago in one of the several incarnations of the *Saturday Evening Post,* a magazine distinguished by the cover art of Norman Rockwell and short stories and commentary by well-known writers and humorists. Admiration for the cormorant comes in a poem published by the popular glossy magazine in 1960.[2]

The Cormorant
The cormorant
With staring eye
With naked beak
And black wing-feather
Walks in the bamboo
Edge of the river
Where tall reeds move
Before his coming
In the last ebb
Of the water's running,
Where the pale shadow
Of the lotus breaks
In the green water,
In the ripple
Of the river,
In the eye
Of the cormorant.
—RUTH ELIZABETH FRAME

The Norse fisherman in the tale of Udrost no doubt encountered cormorants different from our double-cresteds, but nevertheless closely related. Cormorants stepping among the bamboo and exotic lotus blossoms are certainly not the same cormorant species nesting on Little

Galloo Island, but lacking bamboo and lotus, with a few small differences, a cormorant is a cormorant. Most modern fishermen, whether hailing from Scandinavia or Henderson Harbor, would rarely, if ever, feel beholden to a cormorant. Cormorants are, by and large, not well loved by fishermen. However, one exception does occur in Asia, where we see the cormorant not as a spectator to humans hooking and netting fish but employed as the fisherman for humans.

Long before the famed European explorers sought passages to distant lands and continents, Asian fishing masters took advantage of the cormorant's natural fishing ability and its instinct to fish without the aid of line and baited hook. They looked on cormorants as respected, even loved fishing partners rather than adversaries. An archived *New York Times* article dated 1886 tells of an American tourist's observation of an elderly Chinese fisherman working his cormorants, known there as *lu-tse,* from a bamboo raft on a river. The details are appealing, although some phrases may not be considered "politically correct" today.

> Presently he held out his right hand, palm upward toward one of the birds. It waddled briskly and hopped in his open palm. . . . The old native fondly stroked the feathers of the bird on his hand, rubbed his wrinkled cheek along its neck, which he kissed now and then, and talked to it Chinese, evidently in endearing terms. The bird showed much delight at the attentions of its master. It laid its head on the Chinaman's arm, and rubbed its bill against his face.[3]

Since the eighth century, thirteen hundred years ago, Asian fishermen have captured cormorants and trained them to return the fish they caught to their masters. In time, well-treated domesticated birds and entire flocks were revered and passed down through subsequent generations of fishing families. In China, families learned to raise cormorants from eggs laid by the most skillful hunters in their flocks, developing more and more productive strains of captive cormorants.[4]

The traditional method of fishing with cormorants had the bird tethered with a snare or ring around the base of its neck, which allowed it to swallow only small fish while larger fish were lodged in its throat for the fisherman to remove. To concentrate the fish, the fishing master, sometimes working with a number of boats, drove the birds into visible schools of fish, where the cormorants commenced their hunt.

Although mechanized commercial fishing methods have long since replaced the cormorant, the practice continues today on several rivers in China and Japan where modern fishing masters ply the rivers in long, covered riverboats. On these rivers the fisherman's stock and trade is working the cormorants for paying tourists rather than catching fish for market sales. So on the darkest moonless nights they set bright, glowing lanterns or fire grates on the bows of their fishing boats. The lights attract smaller fish to the boat, which in turn attract hungry larger fish for the birds to hunt. When the cormorants return to the boat with fish in their throats the fishermen grill and serve the fresh catch to their touring guests with traditional beers and liquors.

Grilled fish and cold beer, done with nary a hook or bait or line in sight—perhaps a true touch of Udrost's perfection. As we learned earlier, in many situations cormorants were not well-loved, were even hated. Ruth Elizabeth Frame's last line of her poem, "In the eye / Of the cormorant," tells a little bit about the perceived spooky, gothic nature of the cormorant. Its bright eyes, with their mischievous sparkle, broadcast a hint of shrewdness that's frequently construed as a touch of evil or an association with the "dark side." Even a contemporary outdoor writer sailing the length of the Great Lakes as a crew member aboard a seventy-four-ton, tall-masted schooner saw an evil potential in the birds; this time *our* cormorants. "Dozens of double-crested cormorants hunkered on the rocks of the breakwall," he wrote, "their wings held out from their bodies to dry. They seemed vaguely sinister, like rows of cloaked vultures on a parapet." Referring probably to their conflicts with fishermen on the lakes, he added, "Recently, a lot of people have decided that the birds are definitely sinister."[5]

In the other facial feature of the cormorant, the beak or bill with its distinctive curved end, some people also see an evil lilt. The cormorant bill is designed to catch and hold fish, but onshore, on the rocks, and on the pilings, its bill is sometimes portrayed as having a bemused smile, implying knowledge unknown to others, namely humans. But whatever nervous, superstitious people perceive, the cormorant is still only a big black bird that eats fish.

Without a doubt, cormorants do love fish. Some anticormorant activists claim the birds eat several times their weight in fish daily. Other claims include the idea that they eat continuously, ravenously, even angrily and insatiably. That's a lot of judgmental adverbs applied to just one bird. The bird's reported appetite, though exaggerated over time, came to be coined as the adjective *cormorous,* meaning "gluttonous" or "rapacious." And oddly enough, again in the *New York Times,* in 1852, in an editorial titled "Bullying the Government," the author attacked "our foreign neighbors" who used the facilities of the New York Custom House to cheaply move their goods into American markets. These foreign merchants brought incessant suits against the Treasury Department over fees and procedures they thought were unfair. The writer refers to the merchants with the insult, "They are perfect cormorants for profit." It would be very satisfying to think that the reputable *Times* thought of cormorants and foreign merchants as clever, hard workers, but that's probably not what they meant to imply.

It is interesting to note the contrast between how fishermen of old sought cormorants out as guides to the safety of shore and the bounty of an island paradise and how modern fishermen see them as gluttonous, sinister competitors plotting to steal the fish that rightfully belong to the fishermen alone. And yet Chinese families and fishing masters had a distinct fondness for cormorants, the birds that brought food to their tables, with an excess to sell at the market for a modest profit. Perhaps the disparity in attitudes over the ages stems from the huge success cormorants have had in increasing their numbers, even in the face of fierce human persecution.

Regardless of what fishermen, city architects, politicians, and poets did, said, and wrote about cormorants, the fact remains that they evolved; they adapted their way of life to survive and thrive on and in the water, on isolated rock piles like Little Galloo Island, and, as we'll see in the following chapter, in the air.

3

Flight

When a Dinosaur Looked Down upon Gravity

THE CORMORANT is a persistent goal seeker throughout its life cycle. Cormorants do not give up. They adapt. These birds, as examples of adaptive evolution, embody the ability to maximize their use of the environment to survive and prosper.

In modern times, *Phalacrocorax auritus,* the double-crested cormorant, exploits the environmental worlds around it: it swims on the water's surface, dives to its depths, nests and perches on land, and flies. The double-crested cormorant was and continues to be successful because it exploits the opportunities offered in each of its four worlds and copes with the dangers each one poses.

The violence perpetrated on cormorants was often committed out of a *perception* of what was happening rather than the *reality* of actual situations. Early misconceptions of the bird itself caused many of the conflicts. The mistaken beliefs and gothic folklore are to double-crested cormorants what the imagery of Woody Woodpecker, the 1940s zany, red-headed cartoon character with its staccato call, is to the reality of its natural counterpart, *Dryocopus pileatus,* the pileated woodpecker. The one had little relation to the other; the comparison was an overdramatization and exaggeration, used with a specific purpose in mind. Entertainment is entertainment, and its distortions create conflict and drama, which is exciting, but science is science, and distortion and exaggeration create conflicts where none needs exist. It's easy to condemn a species like the cormorant for *apparently* disrupting the ecosystem, but if they are not indeed the culprit then the issues remain unresolved. To value the genuine nature of the cormorant, in this and the next few chapters, we'll look at its niche in today's ecosystem and how it got there.

Sometimes it's easier to understand an animal, particularly an "eccentric" species like the cormorant, by looking at its more conventional relatives and its ancestors. One discussion of the implications of animal evolutionary relationships, what scientists call phylogeny, stated that the "pattern of animal interrelationships has profound consequences for understanding the underlying processes of animal diversity."[1] Also pointed out was the idea that phylogenetic reconstructions, that is, the diagraming of evolutionary pathways or "trees," are noted for their controversies, inconsistencies, and inherent uncertainties. The "trees" are models, works in progress in constant flux, with revisions based on new discoveries welcomed. The same article proposed that even though there is no direct experimental laboratory testing available to duplicate hypotheses of evolutionary history, it is the compilation of independent data and findings that helps classification specialists, called taxonomists, reach reasonable conclusions. The bringing together of independent findings, in the form of fossils, and scientific observations helped produce the probable evolutionary background of birds in general and double-crested cormorants in particular.

When biologists speak of classification and taxonomy what they are really discussing is evolution. In a convoluted way, and with a fair number of exceptions, as a rule of thumb they assume that organisms that resemble each other today most likely followed nearly identical evolutionary paths; they presumably have an ancestor common to each line somewhere in their evolutionary history. So taxonomy, above and beyond the comparison of behaviors and anatomical features, is the possible mapping of an organism's development through the ages.

In the case of the cormorant, it is a vertebrate, in class Aves, the birds. Taxonomists place the cormorant family in the same order, Pelecaniformes, as frigate birds, gannets and boobies, pelicans, snake birds, and tropic birds. The four major families, sorted by numbers of species and individuals, in some ways resemble each other but differ greatly in how each captures its food. Frigate birds use aerial pursuit in their hunt of surface marine animals such as squid, compared to gannets, which capture their prey by means of a dramatic plunge into the water after hovering at a height. White pelicans use their long necks to stretch down into the water to grasp their prey from a wading or swimming position, whereas, like brown pelicans, double-crested cormorants dive to depths from the surface in pursuit of their quarry.

Biological classification is not an exact science and is open to a great deal of interpretation. The conventions of classification dictate that all organisms in a single taxon, or group, share common characteristics, but the number and type of shared features required to constitute a group are indeterminate and decided by the specialist involved in the classification. All six families, or subdivisions, of birds in the pelecaniform order, which again includes cormorants, are linked by two major anatomical characteristics: (1) the colorful, sometimes distensible gular sac or flexible throat patch, located on the neck; and (2) the totipalmate foot.

The gular pouch is the trademark fish-holding pouch for which birds like pelicans are famous. Birds in this group, like the frigate bird, also use the inflated pouch to attract females in their courtship displays. Double-crested cormorants use the patch in courtship displays as well, but cormorants have also evolved a behavior in which they use a gular flutter to evaporate moisture from their respiratory tract. The process creates a radiant-cooling effect to help lower their body temperatures in heat-stressed situations such as those on the cormorant's barren nesting sites.

The totipalmate foot, the second characteristic feature of pelicanlike birds, is different from the webbed foot of ducks and geese with which many of us are familiar. Ducks and geese swim on the surface propelled by feet with three webbed toes forward and an elevated toe on the rear of the leg. The pelecaniform foot instead is distinguished by all four toes pointing forward joined by a single web. The fourth forward toe expands the web, forming a larger paddle with additional surface area, therefore providing more power with each stroke, either when swimming on the surface or in pursuit diving for prey. The increased thrust enables the cormorant to cover more distance in less time on the same volume of air. In short, it makes the double-crested cormorant a more efficient underwater hunter.

Of all the pelecaniforms, the Phalacrocoracidae, the cormorant family, contains the most species. Worldwide, cormorants contribute thirty-eight species to the group, six of which breed in North America. Of these six species, the neotropic, Brant's, pelagic, red-faced, double-crested, and great cormorant, the double-crested is the most common in the United States, the only cormorant to inhabit the Midwest, and generally the most likely cormorant to be seen in the United States. The Brant's, pelagic, and red-faced cormorants inhabit Pacific shores exclusively, and the neotropic is found along the Louisiana and Texas Gulf coasts and in southern Arizona. The great and double-crested cormorants breed along the North Atlantic coast, migrating south in winter, although the great cormorant has moved its breeding grounds farther north in recent years. Our cormorants spend their winters on the southern Atlantic and Gulf coasts from North Carolina to the West Indies and along large inland rivers and lakes northward toward the Midwest.

From the point of view of a taxonomist, the pelecaniform group may actually be more of a classification devised for convenience sake rather than a natural, evolutionary unit, one of the "exceptions" mentioned earlier. This exception concerns how apparently related species may develop similar features through independent evolutionary paths. The assumption, then, that animals that look alike must be biologically linked may be misleading. Organisms that have true shared ancestors are related and said to be homologous, while plants and animals that have evolved along separate but perhaps parallel lines are analogous. Evolutionary biologists refer to the latter type of organization as being the result of convergent evolution, which occurs when similar adaptive features, the totipalmate foot and gular pouch, for instance, appear in different evolutionary branches, producing similar-looking birds that actually may not be closely related at all.

Analogous evolution can occur in the development of an entire organism or only in one or a few of its features. Some organs or tissues may evolve independent of the rest of the animal, particularly when that particular feature has powerful evolutionary implications with great potential for increasing the species' success. Such adaptations include huge evolutionary advances such as the eye, brain, gill, foot, and bird's wing.

Evolutionary changes in cormorants or any other organism are the result of genetic copying errors, mutations, passed along from generation to generation. A possibly more accurate determination of ancestry may be classifications based on changes in DNA. Some evidence indicates that mutations occur at specific, relatively constant rates; therefore the degrees of difference can be correlated with time. In effect, the mutations act as evolutionary clocks that establish sequences of change. Some DNA studies show that the pelecaniforms are not as closely allied as was once thought and suggest that boobies and gannets are directly related to cormorants but tropic birds are unrelated to the others. The pelecaniforms may remain a classification of convenience until further DNA findings provide evidence for a correction.

Again, modern or recent organisms that look alike may do so because they share a common ancestor far back in their evolutionary history. Evolutionary biologists have limited evidence to work with when they attempt to assign these bird families to a single common ancestor. Genetic, morphological, ethological (behavioral), and molecular studies have produced a number of what taxonomists call "sister group" relationships that are not completely understood or completely agreed upon.[2] Sister groups help build working evolutionary models to use as tools until more evidence comes to light. For cormorants the thought, then, is that the totipalmate foot and gular pouch evolved through independent convergent lines resulting in the pelecaniform group, probably the product of several rather than a single common ancestor.

Regardless of internal group relationships, pelecaniforms and cormorants have ancient roots, including similar species perhaps stretching back as survivors from the age of dinosaurs. The oldest pelecaniform fossils include a tropic-type bird found in what is now England from the early Eocene epoch, or about fifty-six million years ago, and a frigate-type bird from present-day Wyoming dated to the same period.[3]

In the movie *Jurassic Park,* aside from the weasely lawyer representing the financial backers, the corrupt system programmer, and the sorely misguided but lovable grandfather, the bad guys seen in so many scenes are the quick, clever, less than lovable, meat-eating velociraptors. Those head-bobbing dinosaurs eat the hunter hunting them, eat most of Samuel L. Jackson, chase the charming children, and make a shambles of the new resort's spotlessly clean stainless steel kitchen. Many scientists now acknowledge that velociraptors or their dinosaur kin turned into the cute little feathered seed eaters crowding around my neighbor's bird feeders and those roasted birds on our holiday tables. How did that happen?

Though not as attractive as cardinals, blue jays, and snowbirds, more realistically, it's easier to visualize the possible dinosaur background of the wild turkeys that also visit the feeders several times a week. They make their first seasonal appearance around Halloween, after the wind and driving rainstorms have stripped the last of the leaves from the oaks behind the house. The turkeys come for the seeds the squirrels have shaken out of the feeders, for the scoops of cracked corn my neighbor sets out for them, and for the huge mast crop of acorns bombarding the ground and pinging off the backyard gas grills. Even though it's not a great time of the year for wild turkeys, with Thanksgiving in the air and hunters in the woods, the flock prowls for easy pickings.

The big old toms strut through the leaves and brush, their long beards swaying in rhythm to their step, scratching the ground for seeds and acorns the gray squirrels haven't eaten or gnawed yet. Being essentially bulky holiday meals with wings, turkeys don't fly often or well, and unless pressed to do so, they usually would rather walk than take to the air. The fall and winter flocks are ordinarily all-male affairs, but hens do show up occasionally for their own easy meal. Females are more prevalent in the spring, when the males puff up, strut hardily, and fan out their broad tail feathers to attract more hens to their harems. And the males are not small birds either. A half dozen of the big toms would probably represent three hundred pounds of "turkey-on-the-hoof," if they had hooves.

The larger males in the flock run the show. The group drifts in the general direction in which the leader heads. When a dispute arises a few quick pecks, a spread of tail feathers, and a kick or two from the leader's spur-armed leg settles it quickly. The turkeys' heads bob and swivel atop their long necks ready to alert their comrades to threatening predators, mostly nosy domestic household pets and snarly coyotes ranging from the nearby mountain ridge. And it's common for at least one of the toms to hop onto the stone fence left over from the land's farming days to gain a better view, patrolling the perimeter for trouble.

Thinking back on *Jurassic Park,* it's not that difficult to imagine turkeys instead of velociraptors strutting through the stainless steel kitchen, heads bobbing, beards swaying, their claws and spurs clicking and clacking and slipping on the polished floor. Instead of stone walls, imagine the tops of stainless steel counters offering a better view. It's also difficult to visualize turkeys twisting a door handle, and who knows what would happen if wild turkeys caught up with the children, but, then again, Samuel L. Jackson's character might have made it into the sequel.

The crux of this turkey scenario is that birds, turkeys as well as more competent flyers like our cormorants, did not materialize out of the blue. Birds as a group developed and emerged from earlier life forms, probably from one or a variety of dinosaurs, which left their marks in the genes, physiology, anatomy, and natural history of their modern-day versions.

Luis M. Chiappe, author of the 2007 beautifully illustrated volume *Glorious Dinosaurs: The Origin and Early Evolution of Birds,* states quite simply, "The origin of birds has always been one of the greatest mysteries of biology" and "given their uniqueness, deciphering the ancestry of birds has been the subject of intense scientific scrutiny since the advent of evolutionary thought."[4] These two very powerful statements place the emergence of birds on a high pedestal, especially considering the range of biological topics and the wide spectrum of evolutionary deliberation, including the more recent descent of the species *Homo sapiens,* humankind itself.

The story of the evolution of flight belongs not just to cormorants but to virtually all birds, including those that have subsequently lost their ability to fly. Flight is an important attribute of cormorants. It signifies the conquering, or at least the exploitation, of one of their four environmental worlds.

Any discussion of the evolution of birds is more than a cold, detached recollection of transitional animals that resulted in a modern class of vertebrates ten thousand species strong. With birds, the story is more than that because it is about leaving the land behind, taking to the air. It's about flight. Like Luis Chiappe, Pat Shipman, in her 1998 book *Taking Wing: "Archaeopteryx" and the Evolution of Bird Flight,* conveys the sense of intrigue associated with studying the origin of flight, writing, "The basic problem—the solving of which makes birds and airplanes so magical—is that of conquering a hostile medium: the air. Gravity is an implacable force and the air a poorly supportive environment."[5] "Magical" marks the point of attraction that has exerted such a pull on so many scientists since shortly after Charles Darwin penned *On the Origin of Species* in 1859. Part of the "magic" in Shipman's pronouncement is the immense time frame involved in the "enormously long history" of birds.[6]

When scientists speak of the evolution of vertebrates they speak in terms of eons, eras, and ages, and they naturally speak in terms of fossils. And when it's bird fossils they are speaking of, they always mention *Archaeopteryx,* the famous fossil bird and probably the most famous fossils ever discovered. Paleontologists have uncovered and described seven specimens of *Archaeopteryx.* The first specimen, found in 1860, is the fossilized impression of a single feather; a single feather about 2½ inches long and ½ inch wide. The specimen from 150 million years earlier is almost indistinguishable in structure from a feather shed from a modern bird. Another sample is a "150-million-year-old *Archaeopteryx lithographia* from the Late Jurassic Solnhofen Limestones of southern Germany—a toothed, crow-sized bird with powerful hand claws and a long bony tail."[7] Early on, critics claimed that the feathers of *Archaeopteryx* were a fraud, etched by hand into the stone matrix in an attempt to bolster Darwin's concepts of evolution. Later it was shown that the details and construction of the fossilized feathers hold true right down to the microscopic level. *Archaeopteryx* is for real.

Nowadays *Archaeopteryx* is considered a type of early bird rather than a transitional form of flying reptile. Its features are too sophisticated and well developed to be an ancestor of the modern bird. It is a bird. The true precursor of birds, an animal with fewer, less specialized avian characteristics, became the target of the search for the common ancestor of modern reptiles and birds.

Most investigators today "agree that birds are living members of theropod dinosaurs."[8] Theropods, meaning "beast footed," are a classification of bipedal, carnivorous dinosaurs that includes the all-time favorite, the giant *Tyrannosaurus rex.* The renowned *T. rex* was in no way an ancestor to the song sparrows nesting in your home's eaves, but its relatives probably were. Small nonflying theropods have been pegged as ancestors of the vertebrate class Aves, the birds.

Theropods, as you would expect, have a number of anatomical features in common with modern birds, such as hollow, thin-walled bones, modifications of their clawed hand and foot digits, but they also had a mouth armed with sharp, recurved teeth for holding and tearing flesh. The obvious anomaly, the teeth, are uncharacteristic of modern birds but were a prominent feature of *Archaeopteryx.*

In recent theories and discussions theropods have taken the day as the designated ancestors of birds, but in that race they weren't always in the lead, and there may yet be an asterisk in the future record. Many renowned investigators became involved in the case for and against the theropod model of the common ancestor. Paleontologists such as Edward Drinker Cope, Othniel Charles Marsh (Cope and Marsh were notorious, competitive "bone collectors" responsible for filling the best-known natural history museums with massive dinosaur fossils), the British anatomist Thomas Henry Huxley, Robert Broom, and of late the Yale University paleontologist John Ostrum had prominent roles in the debate. In the long and complex process of assigning an avian ancestor, other dinosaurs, such as primitive archosauromorphs whose fossils were found in what is now South Africa, also moved in and out of favor.

In their long history, cormorants evolved first as birds, then as more specialized diving birds. One suspected ancestor of diving birds is an animal called *Hesperornis regalis,* a torpedo-shaped, toothless, flightless, foot-propelled diver that lived in warm, shallow, inland seas during the last portion of the "Age of Dinosaurs" between 144 and 65 million years ago. First discovered in 1870, and much larger than the double-crested, *Hesperornis* closely resembled today's cormorants. Its totipalmate foot is a dead ringer for that of our *P. auritus* but may not be *the* cormorant ancestor since "highly specialized foot-propelled divers actually arose several different times quite independently."[9] It was an attractive candidate as an avian forerunner, but *Hesperornis* failed one critical test of ancestry: its line died out seventy to eighty million years ago, while cormorants, with their remarkably similar foot, did not appear in the fossil record until about fifty million years later. *Hesperornis* was extinct long before the cormorant even emerged as a species.

Another clue in the evolution of birds came from a hostile, for the most part unexplored region of the planet. Expeditions consisting of trucks and camels to Mongolia's Gobi Desert in 1922 and 1923 were led by Roy Chapman Andrews, whose persona became the model for the fictional franchised movie character Indiana Jones, right down to the hat. The trek into the Gobi Desert was noted for the first discovery of dinosaur nests, replete with eggs.[10] Since that expedition many other eggs have been extracted from the Gobi. In 1993 members of another expedition conducted by a group from the American Museum of Natural History discovered a partial egg, which, after later reexamination, proved to be that of a type of theropod dinosaur. Most revealing in the fossil was the crystalline structure of the eggshell, the microstructure of which showed "features uniquely shared with birds' eggs."[11] Due primarily to the finds of new fossils and reinterpretations of previously studied specimens, theropods did become the accepted candidate for the avian ancestor—at least until new fossil evidence is uncovered or a fossil long locked away in a museum cabinet is reexamined.

Both ancient and modern birds attained flight because they had wings. Since it is such a complex organ, the wing most likely evolved slowly over long time periods. And like the gular pouch and totipalmate foot, the wing, being so critically important to success and survival, in all likelihood evolved independently several times in several different species. The most widely accepted concept on the evolution of flight, the arboreal theory, was proposed in 1880 by Othniel Charles Marsh, known for his lively and contentious study of dinosaur fossils. Marsh's theory was bolstered and refined in 1965 by Walter Bock of Columbia University. In Bock's version, ground-dwelling quadruped (four-footed) ancestors of birds occasionally took to the trees to search for prey from an elevated vantage point. They later permanently adopted the arboreal way of life and evolved through several stages to become active flyers such as *Archaeopteryx*. These primitive active flyers constituted the pathway to modern birds. Bock's model is shown in the adapted flow chart.[12]

Ground-dwelling ancestor → Adopts arboreal lifestyle → Leaps tree to tree → Plummets → Employs parachuting → Develops gliding → Achieves active flight → Produces modern birds

---

Bock's arboreal hypothesis that "gliding was a precursor to flapping flight" assumed a gradual inclination to an increasing dependence on more efficient flight.[13] The advancement to gliding and parachuting activity may have been two of the easier, less complex changes. These modifications may not have necessarily involved strictly physical adaptations but may have been reached through accumulated changes in behavior and posture. Parachuting and gliding are simple maneuvers in which the animal increases its surface area, slows its descent, and increases its horizontal distance gained, allowing it to land safely. This can be managed by the spreading of arms and legs to dramatically increase the animal's air resistance and drag, which may have been aided later by the evolution of skin folds between digits or the eventual development of flight feathers.[14]

Feathers are modifications of the outer skin unique to birds. Made of a protein called *keratin*, feathers are the most complex appendage of the skin produced by any vertebrate. They may have first evolved as an insulating mechanism for possibly warm-blooded dinosaurs, but the fossil record is unclear. Until recently few if any records of dinosaur fossils showed traces of feathers, but the latest discoveries in China may change what scientists thought they knew.

It's been held as doctrine for some time that feathers first evolved in the two-legged, meat-eating theropods sporting stiff, featherlike structures, so-called dinofuzz.[15] The flightless dinosaur that roamed northeastern China was identified as a dromaeosaur, a relative of the *Jurassic Park* villain velociraptors whose feathers are said by some scientists to have been structurally similar to those of modern birds. Adding to the feather controversy is the fact that the Chinese creature was only very distantly related to carnivorous theropods and was far older than other primitive birds. If the Chinese fossils are interpreted correctly, and the time scales are accurate, then the development of feathers may have taken place on an evolutionary path that was completely independent of the beginnings of avian flight.[16]

In our modern age, birds use two distinctive methods of flight. Some birds, such as eagles, hawks, and vultures, spend much of their time aloft using air currents and updrafts in soaring or gliding flight, with little or no wing flapping. Other birds, like our double-crested cormorants, have wings unsuitable for soaring, so they must flap their wings continuously in flight. Birds exhibiting flapping flight have strong, ultralight flight feathers that provide forward thrust and lift by acting as asymmetrical airfoils whose individual cross sections resemble that of an airplane wing. The primary flight feathers, which allow the bird to gain initial altitude (also the feathers that are trimmed when a pet bird's wings are "clipped"), are attached to the "hand" of the wing while the secondary feathers, which help provide additional lift by shaping the wing into an airfoil, are attached farther up the limb at the "forearm." Similar to flight feathers in construction, the tail feathers extend from the fused caudal vertebrae in the spine and work to provide additional lift and stability in flight.

The feathers of the wingtip also play an important role. In flight, especially in V-shaped formations seen in cormorants, ducks, and other waterbirds, the complicated airflow pattern around the wingtip generates a vortex, driving swirls of air upward and to the rear of each wingtip. The swirling updraft coming off the wingtip of the leading bird in the formation extends for a number of feet behind the bird, which allows the following bird to conserve valuable energy, both in short flights and certainly on long migratory treks.[17]

In an effort to describe a bird's flight efficiency, scientists developed the concept of "wing loading." The calculation is the simple ratio of the bird's weight to the surface area of the wing. A bird like the small barn swallow has low wing loading due to its large wings in relationship to its slight body weight.[18] Cormorants, and many other waterbirds, have rather heavy bodies in relation to their wing area, and therefore have high wing loading, mathematically explaining why cormorants require longer distances for their water-surface takeoffs.

Once early birds had evolved a basic, successful, anatomical model for flight, including sufficient limb modifications to produce an effective wing, the original prototype remained little changed millions of years later in modern birds.

Some ornithologists claim that once you remove the feathers all birds look alike. To a degree it's true. All birds had to undergo similar evolutionary changes to take to the air, so their skeletons are relatively uniform compared to structures in other vertebrate groups. And to succeed in the air, the development of wings alone would not solve all the challenges of flight. Several other structural elements would have to come into play.

One such element was the bird's skeletal system. Animals first attempting flight needed to lighten their load. The robust skeletal structure of their dinosaur predecessors was not at all practical for conquering the sky. Early birds still needed strong, compact frames, but heavy, bulky, dinosaur-type bodies were never designed to get off the ground.

One solution to the weight issue was lighter bones. For instance, the dried skeleton of a frigate bird sporting a seven-foot wingspan actually weighs less than the total weight of its feathers.[19] But the bird's bones still needed to be strong enough to withstand the rigors and stresses of flight, as in the lengthy takeoffs and splashy landings of cormorants. Avian bones, termed pneumatic bones, are semihollow, filled mostly with air and supported internally by crisscrossed structural "struts" for strength. Using these hollows for practical purposes, the bones contain extensions of air sacs, which have direct connections to the rest of the respiratory system. Even though diving birds eventually developed a slightly heavier bone structure to help overcome their buoyancy, air sacs in the pneumatic bones helped provide greater oxygen capacity for longer, more productive dives.

Another solution to the weight versus flight problem was an actual reduction in the number of bones through the deletion and fusion of some skeletal structures. Take, for example, the following modifications.

1. Skull bones became fused into a single unit. Ankle and foot bones fused to form the characteristic avian leg bone.
2. Fragile, separate ribs were joined together with hooked extensions of bone known as uncinate processes, which added structural strength and still created a lightweight, rigid rib cage.
3. The sternum or breastbone was reinforced with a tall keel for the solid attachment of powerful flight muscles.
4. Avian clavicles, or collarbones, were fused together into a single bone—the furcula—the familiar "wishbone" of our holiday turkey.
5. Vertebrae were fused into three separate groups in the spinal column, producing specific functional supports for wing flapping and landing and a structural support for the tail.
6. Only three digits in the "hand" of the wing evolved instead of the typical five-fingered vertebrate model.

In a real and actual sense, the cormorant did conquer its four worlds, but the conquests were not absolute, perfect, or complete. Like many other natural phenomena, the cormorant is a system of compromises. It is only a moderately efficient flyer. It wobbles on land and takes off poorly. And we'll see in the following chapter that the cormorant does dive well—within certain limits. Each of the cormorant's evolutionary threads accomplished specific goals for specific environments. These threads built a bird that may not have unconditionally conquered four worlds but certainly carved out a niche in each of them and found homes in and exploited both fresh- and saltwater ecosystems, above and below the water's surface.

4

A Time to Sink *and* Swim

WHEN A SPECIES manages to occupy a new ecological niche the adjustment signifies that the organism has undergone a number of physical and behavioral adaptations to permit it to do so. But when a species makes the quantum leap to inhabit an entirely new medium, it's a sign that huge changes have taken place in the species' behavior patterns and throughout its body.

Simple plants in the sea adapted to conditions ashore and gained a foothold on land as the result of long-term mutations. When marine animals first stepped onto sandy beaches or climbed onto rocky ledges they changed the way they lived, how they moved, how they breathed, how they gathered food, and what they ate. These transformations took ages to accumulate and continued to build on each other through the immense power of natural selection. As in the cormorant's ancestor's leap into the air, its plunge beneath the water's surface built into the bird's constitution a greater genetic propensity for flexibility. It confirmed the cormorant's presence as an active predator, not a scavenger. But in some ways this same flexibility forced the cormorant into a set of rigidly fixed behavior patterns: with very few exceptions it eats only fish. To feed on fish it must live near water. And as we'll see, the double-crested cormorant's diving range is limited, so it must hunt along shallow shores, preferably those of large bodies of water to ensure a constant variety and supply of prey. Adding the subsurface world to its dining options changed the cormorant into the bird we see today.

Cormorants catch their prey using a method known as pursuit diving. Pursuit-dive foraging is a widespread practice in the avian world, and as many as 150 species across the globe feed in this manner. The small forage fish they feed on thrive in schools, usually swimming adjacent to shorelines, near submerged structures, or close to the bottom to help avoid being ambushed by predators. As expert predators, double-crested cormorants have become adept at shallow-water feeding maneuvers and rarely forage far from shore; a behavior that many times puts the birds in direct conflict with commercial and recreational fishermen.

Any animal that dives from the water's surface to the depths faces a number of challenges from the moment it dives, including a complete change in the very medium through which it travels. The physiological demands that allow a bird to dive are far different than the ones required for flight, walking, or swimming on the surface. In the evolution of flight many of the adaptations that support one type of locomotion may limit the other. It is this conflict of advantages that adds still another layer to the interesting complexity of the double-crested cormorant.

Water, as an environment in which to live even on a part-time basis, has its own unique set of physical characteristics. It is far heavier than atmospheric air and thus produces greater pressures with increased compression of biological tissues and air spaces. Being heavier, water is also denser than air, in fact eight hundred times denser, making the effort to overcome resistance more difficult and requiring more energy to move through it. Water also reflects, absorbs, and bends light waves entering it from the atmosphere, greatly reducing visibility and distorting underwater images.

Even the molecular structure of water works against an animal diving from the surface. It is the same feature that keeps ocean temperatures relatively stable and more slowly affected by temperature changes than air. The hydrogen bonds between oxygen and hydrogen atoms in the H_2O molecule allow a body of water to absorb a great deal of energy before changing temperature itself. Water also has a property that lets it draw heat from a warmer body many times faster than air at the same temperature. In other words, the thermal conductivity of water is about twenty-five times greater than that of air, meaning that an unprotected submerged animal loses its body heat twenty-five times more quickly in water than it does in air, possibly faster than the animal can generate heat by consuming and digesting food. Cormorants, as diving birds, therefore play a tricky balancing game of energy production versus energy conservation in every pursuit dive they perform.

And then there's the matter of oxygen. It's essential. Animals require oxygen to survive. Birds, mammals, reptiles, and some amphibians obtain oxygen directly from the air by means of their open respiratory systems, usually through the lungs. Fish and some other amphibians extract dissolved oxygen from the water surrounding them via gills or similar structures. During their dives cormorants must carry out their underwater chores using only the oxygen they carry within their own bodies. The efficiency of using that oxygen creates the margin between a successful and an unsuccessful dive and perhaps between eating and going hungry.

The differences in the physical properties of air and water thus present all sorts of challenges to cormorants as they dive. Picture a human swimmer, clad in only a pair of swim trunks, deciding to dive into the depths as a free diver, that is, holding his breath, and lacking a portable air supply. When humans put their heads beneath the surface and open their eyes, one of the first things they experience is blurred vision. The lens and other components of the human eye evolved in an atmospheric environment, not under water. The requirements for underwater vision are far different from those needed for seeing in air. For the human eye to function under water it must adapt to the altered characteristics of spectral composition, luminescence, and turbidity in its new environment. And since water is far denser than air, it also creates unique pressures on the eye, which our diver has to overcome in order to see at all well under water.

To partially counteract these distortions and pressure effects the diver places an air space between his eyes and the surrounding water in the form of a face mask with a clear glass plate. The mask permits the eye to function normally at or just below the surface, but as the diver descends water pressure builds against the mask, pressing it more and more tightly against his face, compressing the air space and soon his eyes. Divers call it "eye squeeze," and uncontrolled squeeze can cause permanent damage to a diver's eyes. To neutralize the uncomfortable and eye-damaging pressure the diver simply exhales a small amount of air through his nose into the face mask, quickly equalizing the pressure. In most hunting situations cormorants must see their prey under water in order to pursue and seize it with their bills. Unfortunately, since the face mask is a human invention and not readily available to double-crested cormorants, they had to evolve other tools.

In the cormorant, an important adaptation to increased pressure occurred in the lens of the eye itself and the muscles controlling it. In a free dive, the increased water pressure on an exposed human eye distorts the shape of the lens, spoiling its ability to focus a clear image on the retina. The cormorant lens, unlike the human lens, is highly pliable, having gained the flexibility to accommodate changing conditions and depths. During its evolution the cormorant's strong intraocular muscles developed the capacity to effectively control the shape and curvature of the lens. Thus, in the cormorant's pursuit dive, as the bird descends and the water pressure builds on the eye, the intraocular muscles continually adjust the shape of the lens, which in turn allows the eye to produce a relatively undistorted prey image on the bird's retina.

The ability to focus an image alone does not necessarily guarantee flawless vision. At one time cormorants were thought to have hawklike vision underwater. How else could they catch fast, fleeing fish? There may be an alternative. Several detailed studies of cormorant eyesight came to the conclusion that cormorants may actually have rather limited underwater eyesight. The evidence shows that under low visibility, in situations involving high prey density, cormorants may actually use the sensitivity of their bills in the final moments of the hunt to feel the movements of their nearby quarry and capture it. Other investigations report that cormorants, unlike many other birds living in an air environment, evolved highly mobile eyes, capable of wide, independent ranges of movement or even separate swivel actions to compensate for the reduction in brightness and loss of a particularly sharp retinal image.

So even though cormorants can distinguish some detail under water, their visual fields and powerful eye muscles allow them to hunt much like herons, striking out at nearby mobile prey with their hooked bills, thrust forward by their muscular necks, perhaps without knowing their exact prey species or its size. Limited eyesight in the depths may also explain the occasional death of a cormorant by choking on an oversized spiny fish it was unable to swallow completely or release. Controversy still exists as to how well double-crested cormorants see in or out of water, but there is little doubt, among fishermen, fish farm managers, and everyday observers, about the cormorant's legendary fishing prowess.

If we look back again at human divers, another piece of equipment they use to reach the depths and return to the surface is diving fins. Typically constructed with longitudinal ribs as stiffeners, the diving fin acts as a flexible extension of the diver's foot and provides forward thrust and movement powered by the diver's strong leg muscles. The construction of the diving fin is closely modeled after the diving bird's foot, bringing to mind the cormorant's totipalmate foot. The four toes webbed together as a single unit provide a pattern for the longitudinal ribs in the flexible, rubber structure of the human diver's diving fin. And like the location of the fins on the human prone-positioned diver, the cormorant propels itself by means of its feet, located far back on its streamlined body, providing for quick maneuvering like the stern-mounted rudder of a boat.

The cormorant's foot has also adapted to provide a resistance-reducing forward sweep through the water. In its evolution the foot developed the capacity for lateral compression of the foot bones, allowing the foot to fold to a certain degree, shrinking its underwater profile, thus reducing the turbulence and resistance on the foot's forward stroke.[1] In practical terms, this means that the cormorant foot uses and conserves energy more efficiently, resulting in faster, more productive dives, making it a better hunter of fish.

In continuing our comparison to the human free diver, two other considerations the diver must face have opposing outcomes. Since water quickly draws heat from the body, human divers need insulation from the chilly waters as they dive deeper. Insulation, that fluffy fiberglass insulation used in the home, for example, involves insertion of multiple air spaces between warmth and cold. The air enclosed in the insulation slows the transmission of heat, conserving energy. For this purpose divers don wet suits whose neoprene foam, constructed of chambered air cells, creates an insulating layer between their skin and the heat-absorbing water. And then, to counteract the buoyancy generated by the wet suit, divers also need to strap on a weight belt to gain neutral buoyancy. The added weight allows them to dive easily, even suspend themselves "weightlessly" in the water column, and return safely to the surface, all the while expending the least amount of energy.

As a parallel, the double-crested cormorant uses its feathers in the same way our diver uses a wet suit. Anyone who has ever handled feathers knows how light, airy, and nearly weightless they feel. The structure and layering of feathers in cormorants naturally traps pockets of air close to the body when the bird dives, which, combined with air enclosed in the hollow structure of the feathers themselves, provides an insulating layer similar to our diver's neoprene wrapper. And just like in the wet suit scenario, this trapped air greatly adds to the bird's buoyancy and so at the same time reduces the ease of diving. As with the diving mask, the long evolution of diving birds made no provision for weight belts, and instead diving birds developed somewhat heavier bones than terrestrial birds to provide an increased weight component to their underwater hunts. But building heavier bones was not the entire solution.

An additional approach to improving diving ability is for the bird to become "less light" as opposed to becoming "more heavy." Aquatic birds normally use the fatty, waxy oil secreted from their uropygial gland, or oil gland, located at the base of their tails, to preen, waterproof, and help protect their feathers from annoying and potentially disease-carrying parasites. It's common to see birds transferring preening oil from the oil gland in their tails to their feathers using their heads or bills to spread the oil. The overall result is a waterproof, more buoyant bird. Double-crested cormorants do have an uropygial gland, but it doesn't produce a great deal of oil. As a consequence, cormorants have essentially, "wettable" plumage, which allows water to permeate the feathers, helping to cancel buoyancy effects. Its wettable plumage is the reason why, after diving, cormorants frequently perch at sites where they can expand their wings into the breeze in a spread-eagle fashion to dry their wet feathers.

Aside from the buoyant effects of feathers, diving birds also needed to overcome the consequence of their internal flotation: the many air sacs connected to their respiratory systems. To do this, double-crested cormorants, and all other diving birds, exploit the natural relationship between gasses and pressures. And as with insulation and leg placement, this relationship works for and against the very diving efficiency that diving birds attempted to achieve in their evolution.

The physical relationship between pressure and gas is expressed in Boyle's law, a principle familiar to high school physics students. It states that under steady temperature and quantity conditions, there is an inverse relation between the volume and pressure of a gas. In plain language, it means that any gas or mixtures of gases, including oxygen and air, will compress proportionately to the amount of pressure exerted on it: the more pressure, the less (smaller volume of) gas.

From a diving bird's practical point of view, Boyle's law suggests that the deeper it dives the more its body, air spaces, and oxygen are compressed. In fact, in a descent of just thirty-three feet into the water column the hydrostatic pressure doubles, compressing the gasses to half their original volume. At sixty-six feet, the pressure triples, reducing gas volume to one-third of what it was at the surface. For our cormorants this translates into an easier initial diving effort. After beginning the plunge, when the bird first overcomes its surface buoyancy, gasses within its body immediately begin to compress and become denser and heavier, thus decreasing the bird's natural buoyancy with every unit of depth it gains. The deeper it dives the easier it is to dive deeper—as long as its oxygen supply holds out. It is at this point that Boyle's law begins to work against the dive.

Animals rely on oxygen for the body's chemical reactions to produce energy. In air-breathing animals, oxygen, comprising about 20 percent by volume in the atmosphere, is extracted by the lungs and absorbed into the bloodstream for use in the body's tissues, such as the muscles. As the surrounding pressure continues to build, gasses in the lungs compress to a still smaller volume, reducing the practical amount of oxygen available to the respiratory and circulatory systems and consequently reducing the amount of energy produced in the tissues. If this process is not offset in some way the dive time is foreshortened, forcing the cormorant to the surface with no food to show for its energy-expending dive.

In experimental studies on captive diving birds, including cormorants, scientists discovered that many species have the capacity to dive for longer times than their calculated oxygen (O_2) volume would normally allow. Something else was in play. Once again, in their development and conquest of the underwater world, cormorants evolved additional mechanisms to counteract compression effects that would otherwise limit the duration and efficiency of their dives. As a result of these changes a modern diving bird stores oxygen in three separate areas in its body: the respiratory system, the circulatory system, and its skeletal muscles.

If we were charged with the task of building a catalog of ideal adaptations for diving birds, we would first probably add larger lungs to collect and store more oxygen, but nature sometimes works contrary to intuitive thinking. Larger lungs may store more air but not make the respiratory system more efficient. In vertebrates such as mammals, air enters the lungs where a certain amount of oxygen is extracted by the air sacs, or alveoli, and transferred to the animal's oxygen-carrying red blood cells. At the same point, carbon dioxide, CO_2, a byproduct of cellular metabolism and exertion, diffuses back into the lungs from the alveoli and is exhaled. The mammalian system seems to work well and at first seems concise and tidy, but in its own way it is inefficient, since oxygen-laden fresh air entering the lungs mixes with the carbon dioxide waste being exhaled. A mammal's lungs act as a sort of cul de sac, or "dead end," in its respiratory system. Not all the available oxygen is absorbed, and not all the CO_2 is expelled with each breath. Therefore, a percentage of fresh air inhaled is exhaled, unchanged and unused. A larger lung would not resolve the mixing problem.

Birds, it turns out, have smaller rather than larger lungs in proportion to their body size than most vertebrates. To accommodate the high-energy requirements of diving, as well as flight, birds evolved a unidirectional "flow-through" respiratory system in which the lungs can be ventilated almost completely at each breath, as opposed to the "dead end," mixed-gas scheme of other vertebrates.

In diving birds, such as double-crested cormorants, the air entering the lungs through the mouth and divided trachea (windpipe) passes through the lungs into smaller and smaller passages or bronchi. It is in the finely divided bronchi that oxygen diffuses into air capillaries surrounded by networks of blood capillaries. Carbon dioxide collected in the capillaries passes into the air sacs and is eventually expelled through the trachea and nostrils, bypassing the lungs during exhalation. Therefore, air contaminated with carbon dioxide never returns to or mixes with the fresh air entering the lungs. This rapid, nonmixing, more efficient exchange of gases helps fuel the cormorant's most energetic activities such as pursuit dives and flight.

In the second storage center for oxygen in diving birds, the circulatory system, O_2 is found combined with hemoglobin (Hb), the complex, iron-containing protein found in vertebrate red blood cells. Studies have shown that birds using pursuit diving foraging have a proportionately larger blood volume and a higher concentration of Hb, greatly increasing the cormorant's O_2 carrying capacity and its potential dive time. The cormorant's increased blood volume also serves a purpose in the buoyancy issue. The buoyancy of oxygen stored as a gas in air sacs has a far greater flotation effect than the buoyancy of oxygen stored in a chemical bond in a fluid such as blood. Therefore, the evolutionary increase in blood volume, instead of air sac volume, produces a heavier submerged bird, making for easier, less-energy-consuming dives.

The third reservoir for oxygen in diving birds, the skeletal muscles, works through a small, bright red, iron-containing protein called myoglobin (Mb) found in animal muscle cells. It's the substance that gives red meat its color. Myoglobin is the oxygen-storage unit that supplies O_2 to working muscles, but it is released only when the hemoglobin in the animal's circulating blood can no longer satisfy the oxygen demands of its active muscles. So, as the cormorant dives, its oxygen reserves shrink under increasing water-column pressure, and as its muscle tissue receives less and less oxygen from hemoglobin in the circulatory system, oxygen-rich myoglobin is released providing the additional O_2 necessary to energize working muscles and hence extend the dive. And as in the progression of other avian adaptations, diving birds added one other step. To maximize the efficiency of this system, the skeletal muscles of diving birds typically contain up to ten times higher Mb concentrations than those found in the same muscle types of nondiving birds. But one would expect nothing less from our resourceful cormorants.

So far we have identified two physiological keys to successful underwater pursuit: oxygen and energy. These two essential elements determine the possible duration of a dive, which influences the depth and range the cormorant can explore as it forages for prey.

The production of energy in most animals, including birds, a process called cellular respiration, is accomplished through a complex chain of chemical reactions that take place within living cells. The process includes three major steps: glycolysis, the Krebs cycle, and the electron transport chain. Each step in turn feeds its end products into the next step, building energy-laden molecules.

Glycolysis takes place within the cytoplasm of the cell and breaks large glucose molecules into smaller sugars used in the Krebs cycle. This step, and the electron transport chain, takes place in distinct, rod-shaped organelles or reaction sites called mitochondria, which float in the cytoplasm of all body cells. It is in the mitochondrion that the smaller sugar is chemically oxidized, or combined with oxygen in the Krebs cycle, and then used in the electron transfer chain. The end products of the reactions are water, carbon dioxide, and a compound containing energy-laden phosphate bonds called ATP. The cell then draws on the stored energy in ATP to carry out its specific function, such as the act of contracting in muscle cells, in the case of the double-crested cormorant, the contraction of the skeletal muscle cells responsible for locomotion. The process of cellular respiration is summarized in the expression below.

$$C_6H_{12}O_6 + 6O_2 \rightarrow 6CO_2 + 6H_2O + ATP$$
(Glucose + Oxygen YIELDS Carbon Dioxide + Water + Energy)

One big problem arises for cormorants when their O_2 is depleted and the bird has not yet finished its underwater hunting chores. Hemoglobin and myoglobin have both done their jobs, but the bird is still submerged with no oxygen to feed its cellular respiration. And unlike other diving animals, including humans, who experience a "bursting" pressure and tension in their chests and lungs, cormorants have no physiological alarms to warn them that they are running out of oxygen. The bird consequently relies on anaerobic respiration, a shortcut process of breaking down glucose and completing energy building in the *absence* of oxygen. Found in all vertebrates, it is especially important in diving animals such as whales and seals and in cormorants and other diving birds. Anaerobic respiration produces only a fraction of the energy of cellular respiration and comes into play when the diver's O_2 reserves are exhausted. Even though it is a tool for sustaining longer dives, it is very inefficient, results in an oxygen deficit, and one of its end products, lactic acid, produces tremendous muscle fatigue as it accumulates in the animal's tissues. Lactic acid buildup in muscles is the same "burning" athletes experience when competing in long-distance or intense physical events. The trade-off here, somewhat like that in human divers if they overextend their stay below the surface, is that the cormorant must rest for a longer time on the surface before its next dive, restoring the body's O_2 content and purging the lactic acid from its tissues.

Evolution made one more addition to the cormorant's menu of diving options. During the exertion of diving, oxygen and energy are consumed by various tissues throughout the cormorant's body, but some of these tissues, and the organs they comprise, may not be directly involved with the task of pursuit diving, so the efficiency of their jobs may temporarily be sacrificed for the success of the dive. Physiologically, these tissues may be called on to function at lower, slower levels. This is accomplished through a reduction of blood flow by constricting blood vessels that supply organs not directly involved in the dive, as long as the organs can tolerate the reduced O_2 levels. Under conditions of extreme physiological stress, regulating mechanisms may actually induce temporary localized hypothermia, reduced working temperatures, in tolerant tissues that are also not being called on to perform at the time.

Scientists have identified and classified about 1.5 million species of animals on Earth. Of those 1,500,000 species, only about 150, or one one-hundredth of 1 percent, can do what cormorants do. Overcoming the factors of hydrostatic pressure, buoyancy, Boyle's law, blood volume, energy production and conservation, the limitations of hemoglobin and myoglobin, and all the others, the double-crested cormorant represents more than an accumulation of fifty million years of evolutionary adaptations. They conquered four separate environmental worlds and by doing so showed themselves to be formidable human competitors. Cormorants are adaptive, resourceful animals capable as a species of surviving predators and storms, the ire of fishermen and fish farmers, and the marksmanship of armed government "technicians." But the species was not and is not immune to everything in its environment.

Unfortunately, the qualities of persistence, adaptability, and a powerful reproductive drive were ineffective weapons against the greatest challenge cormorants would face: the discovery and widespread application of the chemical compound dichloro-diphenyl-trichloroethane, the persistent insecticide known as DDT.

5

The Face of Extinction
Eggshells versus DDT

EARLIER WE EXPLORED how birds like the double-crested cormorant most likely evolved and how they faced the many challenges in their four worlds. But there is a difference between challenges that tested the vigor of individuals and those that attacked the entire species. The cormorant's flexibility, perseverance, and tenacious breeding behavior kept the species a step or two ahead of its natural predators, the hostile environment, and human hunters. These agents attacked and killed birds in ones and twos and sometimes birds in the hundreds, but the larger population flourished because the cormorant's vast numbers, scattered in various flocks and colonies, provided a buffer to protect the species. But what this species, *Phalacrocorax auritus,* could not possibly endure was a prolonged, pronounced attack on its capacity to simply lay viable eggs and produce healthy chicks. Cormorants had yet to confront the massive release into the environment of the class of manmade chemicals called organochlorines, the most recognized of which is the notorious DDT.

Cormorant eggs, the eggs we see in other birds' nests, and even the common chicken eggs in cartons in our markets are really nothing more than specialized ova, or oocytes: egg cells. Birds' eggs, though covered with protective shells, are similar to those found throughout the animal kingdom, including those of humans. Controlled by hormones secreted into the bloodstream, the female's ovary releases the ovum, functionally considered a yolk and nutrient source at this point, into the oviduct where it may or may not be fertilized by the male's sperm. As it moves through the oviduct, various layers, such as membranes, structural fibers, and albumins, are added to build the familiar bird's egg of a yolk surrounded by the "whites." Just before the egg is laid, the shell, formed of calcium salts drawn from the female's bloodstream, is deposited around the egg in the lower region of the oviduct. In a fertilized egg, the blastodisc, the spot on the yolk where the fertilizing sperm entered the egg, then develops into the embryo, protected by the calcite shell. Scientists describe the shell as "a mediating boundary that operates along with the nest microenvironment and the behavior of the brooding parent(s) to isolate the embryo from the external environment."[1] After incubation by one or both brooding parents, and a series of rotations in the nest, the chick breaks through the shell, entering the outside world.

During the incubation period the shell encloses the future chick's entire world and protects it from predators and destructive invertebrates. And in some birds, specific classes of proteins associated with their eggshells have been shown to have antimicrobial activity, thus also protecting the enclosed embryo from harmful invading bacteria common in the parents' everyday environment.

The shell itself is a very porous guardian of the embryo. A domestic chicken egg, for example, has about 7,500 pores, located mostly on the blunt or rounded end of the shell. The pores provide a safe biological link to the outside. Respiratory gases, expelled carbon dioxide and incoming oxygen, pass readily through the pores to satisfy the embryo's metabolic requirements. The texture of the shell varies with the species. Many birds lay smooth, glossy eggs, ducks lay oily waterproof eggs, and double-crested cormorants lay clutches of rough, chalky eggs. The varied textures of eggshells are actually determined by the deposition of the cuticle, an external protein layer that also adds structural strength to the shell.[2]

A bird's eggshell is both fragile and strong, an assembly composed of a combination of calcium and magnesium salts distributed throughout the fibrous matrix of the shell. Calcite, the major component of eggshells, is a carbonate mineral, the same hard material found in the familiar durable shells of clams and oysters we find on the beach. The eggshell is important to the developing chick in other ways as well.

During its development, the bird embryo transfers these same calcium salts from the eggshell to aid in building its own growing bones. So it's easy to see how important the biological regulation of calcium is in the reproductive cycle of the female bird. It follows that any interference with the female bird's calcium intake, production, and deposition sequence would disrupt the protection and development of the embryo within the egg.

On its own, Mother Nature never produced an environmental agent to tinker with a bird's ability to reproduce. That took the hand of man. And in that hand was the agent DDT, dichloro-diphenyl-trichoroethane, the most famous, or infamous, of the organochlorines. These compounds are referred to as organic chemicals, classed as organic because a ring or chain of carbon atoms forms the core of their structure but never originating as a result of natural biological functions. Chlorine and hydrogen atoms make up the rest of the molecule.

As a totally synthetic compound, DDT was created in 1874 by an Austrian chemist, Othmar Ziedler, working in a lab at the University of Strasbourg in Alsace, France. Some accounts report that it was discovered in a search for military chemical weapons. It was not until sixty-five years later, in 1939, that a Swiss chemist and skilled technologist, Paul Hermann Muller, at Geigy Pharmaceutical, while searching for a durable pesticide to use against the clothes moth, uncovered DDT's effectiveness as a contact poison against a wide range of insects.[3] As an insecticide, it was referred to as the "atomic bomb" of pesticides and the "miracle pesticide." It was soon used in public health campaigns against the insect vectors of malaria, typhus, sleeping sickness, and yellow fever. And in agricultural areas it was applied in the field to reduce or eliminate insect pests in the soil that attacked valuable standing crops. Muller's work was of such massive biological, and of course economic, importance that in 1948 he received the Nobel Prize for Physiology and Medicine.

As a pure compound, DDT is a colorless crystalline solid at room temperature. In the laboratory, chemists synthesized the manmade compound from the reaction of trichloromethanol with chlorobenzene using sulfuric acid as a chemical catalyst, a meeting of compounds that would never take place in nature. These precursor chemicals themselves are dangerous to the environment and are components of other deadly compounds such as phosgene, the feared poison gas of World War I.

The chemical and physical properties of DDT are what make it such a useful and dangerous compound. Like many other organic compounds, DDT is chemically hydrophobic, meaning it is nearly insoluble in water. But, also as an organic compound, it is very soluble in organic solvents and also soluble in fats and oils such as those that comprise living animal tissues, including the liver, the brain, and the nervous system. On the molecular level, the chemical bonds between carbon and chloride atoms are extremely strong, producing a very stable, persistent compound not readily broken down in normal environmental situations. So once DDT is introduced into a system it does not dissolve in rainwater, evaporate into the air, or merge readily with gasses already in the atmosphere; it is more likely to linger for a long time.

The laboratory varieties of DDT we find in the real world are known primarily for their marketable qualities. Commercially, DDT appeared as a component of many trade name products, including Anofex, Dicophane, Dinocide, Gesarol, Ixeodex, and Koposol.[4] Designed for a number of public health and agricultural applications, DDT became available as a waxy solid or colorless crystals. Most popular were the prepared formulations ready to use as aerosols, wettable and dustable powders, granules, and emulsifiable concentrates. It was typically sold to final users as a powder to be applied as a dust or in aqueous suspensions.[5] Sometimes referred to as "total DDT," the commercial compounds, typically composed of 65 to 80 percent active ingredients, were actually blends of several related chemicals and isomers. Two of the related compounds, DDE and DDD, also biological breakdown products of DDT, proved to be major culprits in the attack on cormorant reproduction.

Quickly DDT became the insecticide of choice. It was touted as a powerful compound that killed a wide range of insect pests and, when applied in recommended concentrations, was harmless to humans as well as wildlife. In the 1940s, during the latter days of World War II, when the Allies were on the offensive against Germany and Japan, the US military adopted DDT as a way to protect its personnel from insect-carried diseases in the field. The organochlorine was sprayed and sprinkled directly on troops, refugees, and displaced persons as a potent powder in an attempt to control typhus. Typhus is a feared, deadly, highly contagious infection transmitted by lice and caused by bacteria that invade and destroy living human cells.[6] Historically, typhus epidemics wiped out prison populations, cities under siege, and attacking armies; the Russian army lost three million soldiers to typhus epidemics in World War I. Any crowded human population was a potential living culture medium for the typhus bacterium. Victims of this terrible disease suffered fiery hot fevers, delirium, nausea, vomiting, and gangrenous sores. In Italy during World War II, entire cities were dusted from the air with DDT to fight the infection. As a result of these massive campaigns, DDT applications virtually eliminated typhus in many parts of Europe.

In addition to typhus, in the South Pacific, as US forces took back isolated jungle islands such as Guadalcanal and Iwo Jima from the Japanese, another debilitating disease, malaria, struck down as many or more US troops, through fever, fatigue, and system failures, than Japanese bullets. Some units suffered 100 percent casualties due to malaria. Gen. Douglas MacArthur was credited with saying that he knew he could defeat the Japanese but he wasn't sure about defeating the *Anopheles* mosquito, the insect that transmits the malaria parasite: "This will be a long war if for every division I have facing the enemy I must count on a second division in hospital with malaria and a third division convalescing form this debilitating disease."[7] A regular program of aerially spraying the jungles with DDT quickly killed millions of mosquitoes, reducing the rates of new and relapsing malaria cases and keeping more men active, armed, and fighting the enemy. Soon DDT became a friend of both soldiers and sailors ashore. In 1943, it was a staple item included in official US Army supply lists. Millions of servicemen carried cans of DDT to guard against lice, bedbugs, and the ever-present mosquito.[8]

The *Anopheles* mosquito finally succumbed to DDT. As an insecticide, DDT was found to be effective in killing more types and species of insects than any pesticide that preceded it. Its persistence in the environment, later found to be detrimental to animals other than insects, enhanced its potency as an insecticide since it was no longer necessary to spray the chemical directly on the insect; DDT residue could be absorbed by the insect's tissues and so was lethal to the touch. In insects and other animals, the electrical transmission of nerve impulses depends on the regulation of sodium ions and their interaction with nerve cell membranes in a process called the sodium channel. DDT acts at these cell sites, causing the nerve cells to fire randomly and erratically at high rates. As a result, the insect undergoes spasms and paralysis and then dies kicking and twitching. It's easy to see how DDT became MacArthur's weapon of choice against the *Anopheles* mosquito.

After the proven, even spectacular, success the military had in using DDT, in 1945, after wartime federal control was lifted from it, the insecticide was made available to farmers. It was used virtually every place in the United States where crops were grown and forests were maintained and protected. DDT was the farmer's and forester's savior, their wonder drug, their penicillin. It was billed as a cheap, available, potent, and safe solution to farmers' insect losses. And because of its stability and continued toxicity, the residue left in the soil or on leaves and stalks and fruit, just one spraying might protect fields for several seasons. The pesticide was particularly effective in protecting economically important crops by killing voracious insects such as potato beetles, pea aphids, apple-attacking coddling beetles, cabbage caterpillars, corn earworms, cotton bollworms, onion thrips, and tobacco budworms.[9] During the same period, the 1940s–50s, there was little or no control or regulation regarding the manufacture and use of pesticides. In 1947 demand for DDT far exceeded output, even in the face of a 22 percent increase over the previous year's production.[10] Lack of oversight and increasing high demand led to the distribution of impure products, carelessness in handling, and major overapplications.

Before long, US agriculture became heavily dependent on DDT to make farms profitable. This one chemical helped put more food on American tables than was ever seen before. Several agricultural estimates claim that without the application of chemical insecticides insects would devour 30 percent of America's protein supply and 80 percent of the country's high-vitamin crops. The government's, the farmer's, and the public's attitude toward DDT was "How did we ever live without it?"

It is then, in the 1950s, that we see the production, sale, and use of DDT become a tremendous national and global enterprise. The figures are staggering. Worldwide, manufacturers of DDT reached their highest output of 386 million pounds in 1970. In the two decades between 1950 and 1970, the then Soviet Union alone used more than 22,200 tons of DDT *annually*.[11] In the twenty years between 1950 and 1970, the yearly production of the insecticide increased from 125 million to 600 million pounds.[12] The greatest amount of spraying of DDT in the

United States took place in 1959, when 80 million pounds of the pesticide were applied to American fields and forests.[13] DDT was everywhere in our world. One of the most appalling situations involved an agreement between the corporate executives of the Montrose Chemical Corporation of California and local Los Angeles officials. After the Japanese surrender ended World War II, together the corporation and the city developed a plan to discharge hundreds of metric tons of minimally treated DDT waste into the county's sewers, a system that eventually flowed into the Pacific Ocean. It later came to be known as the Montrose Superfund Site on the federal government's National Priorities List of hazardous waste sites.[14]

As its use grew so did DDT's reputation. Endorsements came from the US government, chemical manufacturers, and public health departments. In her 1962 best-selling book, *Silent Spring*, Rachel Carson described the world's complete acceptance of the chemical: "DDT is now so universally used that in most minds the product takes on the harmless aspect of the familiar."[15] DDT was an undeniable success.

The insecticide continued to be a huge success until it wasn't such a huge success anymore. Things were changing. Transformations were taking place in the environment that were not so easily explained away. The changes were not as mysterious as the unsubstantiated UFO sightings claimed by a few individuals but were everyday observations made by everyday people. People saw robins and other songbirds fall dead and dying out of trees. Farmers around the world noted that some new generations of insects no longer kicked and twitched themselves to death when sprayed with DDT. Birdwatchers noticed that there just weren't as many eagles, cormorants, and brown pelicans flying overhead. Investigators checking concentrations of pesticides, especially DDT, in ponds, lakes, rivers, oceans, and fish and other wildlife found levels many times higher than the *safe* levels originally applied to fields and crops by local farmers. Some classes of animals, from snails to fish and birds, carried sizable insecticide burdens, experienced gender variations, exhibited substantial reproductive failure, and showed increased rates of birth defects. The early declarations concerning the efficiency of DDT, which included absolute claims of safety for humans and wildlife, were coming under intense scrutiny. The original safety claims were still fundamentally valid if long-term and cumulative effects were ignored. The early laboratory results looked fine on paper, but they did not have a huge correlation with what was happening in the real world. The success of DDT was about to be challenged.

Much like the miner's canary and its mortal sensitivity to gases such as carbon monoxide and methane, one of the first indicators that things were going wrong was the death of so many birds in areas recently or repeatedly treated with DDT. The common robin, for example, which feeds almost exclusively on earthworms, a seemingly harmless occupation, was one of the first easily observed victims of DDT. Beginning in 1954, in an effort to protect its stands of elm trees from the spreading waves of Dutch elm disease, the state of Michigan undertook a spraying program to eradicate populations of elm bark beetles. The beetles carried and transmitted a mircofungus that attacked the xylem or transport tissues within the elm's trunk and eventually killed the tree. Many of the state's elms provided shade for students and visitors to Michigan's university campuses, so these were areas where much of the spraying was centered. The following spring, returning robins settled on the campuses and soon died—by the hundreds and even thousands. University and outside researchers finally uncovered the deadly pathway responsible for killing the robins. The DDT residue settled to the ground in the autumn along with the elm's deciduous leaves. The fallen leaves and their coatings of insecticide then acted as fodder for the millions of earthworms living in the lawn below. The persistent DDT accumulated in the earthworm's gut and other tissues over the winter and was soon devoured by the returning spring robins. The bioaccumulated poison quickly saturated the robins' ovaries, testes, and other tissues at concentrations far higher than what was originally applied to the elms. More than 80 percent of the robins in the sprayed areas died as a defined challenge to the success of dichloro-diphenyl-trichoroethane.[16]

Many other challenges to DDT appeared as similar deaths of birds occurred across the country, including in Oregon and California, where in 1960 hundreds of herons, pelicans, gulls, and other fish-eating birds in wildlife refuges adjacent to DDT-sprayed farms were found dead and dying from insecticide poisoning. Even the lakes within the refuges were found to be laden with insecticides and their breakdown products, DDE and DDD. Plankton and fish in the refuge lakes contained similar contaminants, which had accumulated in the dead and dying birds in a biological sequence reminiscent of what happened to the Michigan robins.[17]

At about the same time as the use of DDT in health programs and agricultural fields expanded to so many parts of the world, public health officials and farmers noticed that some insect species thought once to be virtually eradicated were springing back in unbelievable numbers. The first indication of DDT resistance in insects was discovered shortly after World War II in 1946, when Swedish scientists uncovered populations of DDT-resistant houseflies—before Paul Hermann Muller had even received his Nobel Prize. Soon after, reports of many other instances of insect resistance to DDT's lethality appeared.

The most disturbing reports involved the reemergence of disease-carrying insects like the *Anopheles* mosquito, which first bloomed again in local populations and then spread regionally, nationally, and globally. The sodium-channel, neural-killing mechanism of DDT and related pesticides now worked on fewer and fewer insect species. The similarities in the way many insecticides functioned also produced cross-resistant insect populations in which resistance to one chemical agent brought about resistance to an entire class of insect poisons. The insects responded in their own way to the spectacular success of the "miracle pesticide," just as they had responded for millions of years to other threats to their existence. They evolved.

Through almost countless generations of their short lives, measured usually in days, the mosquito rebounded through the tried-and-true model of the natural selection of mutations. Researchers at the University of Illinois recently discovered how mosquitoes, and possibly other insects, developed a resistance to DDT. They found that under continual attack by insecticides, many times occurring as an adjunct to massive agricultural applications, some *Anopheles* species acquired the ability to metabolize DDT molecules. In simple terms, in DDT-resistant mosquitoes, the gene that protected the mosquito against naturally occurring plant toxins now produced a modified protein, cryptically called CYP6Z1, which, through its new molecular orientation, combined with a single DDT molecule through the addition of oxygen. The DDT was oxidized, disabled, and detoxified, undoing its ability to kill the insect through the sodium-ion channels in the nerve cell membrane.[18] In effect, the resistant mosquito "digested" DDT and so survived its exposure to the insecticide, living on to produce a subsequent generation of genetically resistant disease carriers. Through a DDT-triggered ancestral genetic response, the lowly *Anopheles,* once conquered by General

MacArthur and vanquished by the chemical companies, on its own challenged and defeated the "atomic bomb" of all pesticides.

Changes were also happening in organisms outside the realm of insects and disease carriers. DDT resistance was not the only problem. Genders of some species were less and less defined; males of predatory species in high-insecticide-use areas were losing their "maleness," becoming feminized.[19] In the wild, some male Florida alligators, exposed to high levels of DDT through the food chain, developed deformed genitalia and feminized body features. Many species of marine snails, which are typically both male and female at some period in their life cycle, were now "male only," causing local near extinctions of some species. Even as recently as 2006, a large percentage of Potomac River smallmouth and largemouth bass, and possibly other fish species, were classified as intersex, meaning that many genetic males were found to be bearing immature eggs within their sex organs.[20] Feminizing processes also took place in a number of gull species in California, Washington, and Idaho and on the Great Lakes. In California, some gull colonies experienced such high rates of feminization in their males that many would not or could not reproduce and left the breeding colony, abandoning a huge excess of lonely breeding females.[21]

In addition to its lethality to insects, DDT exhibits biological properties mimicking animal hormones. Hormones are powerful substances in the body that control and regulate activities of specific target organs, tissues, or cells. In regulating processes such as growth and reproduction, hormones bind to receptors in or on the target cells as a direct function of the very complex configuration of the hormone molecule; it links with the receptor as a key fits with the tumblers of a lock. And because of its specificity, a small quantity of a hormone does a lot of work.

Along with several other classes of contaminants, DDT, DDE, and DDD enter this scenario because portions of their structures closely resemble the *keyed* configuration of hormones, particularly estrogen, the female sex hormone. The chemical pollutants can act in several different ways. The cells may read the synthetic substances as natural hormones and react as though they were, creating inappropriate or untimely physiological reactions. The contaminants may also block the receptor sites on or within cells from attachment by natural hormones, preventing appropriate physiological reactions from occurring. Or the hormone imitators may coax target cells into creating an overabundance of receptor sites capable of receiving many hormone contacts, thus producing disproportionate rates of cellular activity. These synthetic hormones, which circulate through the environment, are absorbed or ingested and go on to initiate or shut down physiological sequences at the wrong times, disrupting biological cycles. The terms describing these estrogenic "foreign" chemicals would comprise a lengthy list. We find terms such as *endocrine disruptors, environmental estrogens,* and the less precise *hormone copycats* and *gender benders* appearing in scientific and popular literature.

With all this information before us about the various natures, properties, and functions of DDT, and its perceived wonderments and evils, we have to ask at this point what it meant to our matter at hand: the double-crested cormorant. What it meant to cormorants was that their numbers plummeted. DDT and its metabolite cousins contaminated the environment to the point of driving the cormorant's reproductive capacity to near zero.

For hundreds of years cormorants had nested in northwestern Ontario in Lake of the Woods, but modern breeding groups did not colonize the Great Lakes until 1913. The birds were not accepted with open arms by fishermen and other local residents but nevertheless managed to expand their foothold. Their numbers continued to soar through the 1920s, 1930s, 1940s, and into 1950s, when breeding cormorant pairs on the Great Lakes totaled in the thousands. Cormorant density reached the level where sanctioned control measures were enacted, primarily in Canada. Unsanctioned acts occurred on both sides of the border. And parallel to what happened with public health mosquito-control programs and agricultural spraying campaigns, conditions in the environment began to change. Cormorants disappeared from waterways and the skies. Even after cormorant control measures, which slowed but did not halt cormorant growth numbers, were discontinued, populations continued to decline from the 1950s through the early 1970s. In 1970, a scant eighty-nine cormorant nests across the Great Lakes were reported by the US Fish and Wildlife Service.[22] And by 1973 cormorant numbers had been reduced by a disturbing 86 percent. In Michigan, the state's Department of Natural Resources placed the double-crested cormorant on its Endangered Species List, classifying it as "probably extirpated" in 1976.[23] Extirpated? What happened to all those birds?

What happened was that the cormorants' veritable ability to reproduce was attacked and weakened. And it took place on several fronts. Numerous studies, conducted on several fish-eating bird species, showed that DDT had accumulated, at amplification factors of "1,000 to 1,000,000 in various aquatic species," and "bioaccumulation may occur in some species at very low environmental concentrations."[24] These highly contaminated fish were in turn eaten by cormorants and other birds. Since bioaccumulation is really an issue of biomass, a simple example illustrates the concept. A cormorant that lives for six years and consumes a pound of fish a day will devour about 2,190 pounds of fish in its lifetime. If we assume that 100 percent of the DDT and DDE in the fish is retained by the bird (not exactly true since some DDT/DDE is excreted in the bird's droppings), then the bird carries a toxic burden equivalent to more than a ton of contaminated fish. That is how DDT jumps from *safe* levels when it's sprayed on a farm to toxic levels in a bird like the cormorant.

In these contaminated birds many studies showed that DDT's estrogenic properties were also magnified, wreaking havoc on the cormorant's DDT-sensitive reproductive processes. The sequence of an oocyte being transformed into the familiar bird's egg requires several "interventions" by hormones to smoothly complete the progression. In high concentrations, DDT, mimicking natural hormones through some or all of these mechanisms, went to work tinkering with the egg's development. The most reported effect was the thinning of eggshells in fish-eating birds; cormorants, bald eagles, brown and white pelicans, ospreys, and others were therefore threatened. Their eggshells became so membranous that the very weight and pressure of the parent trying to incubate an egg crushed it, immediately destroying the embryo within. One study, published by Environment Canada, measured the percentage of eggshell thinning of cormorant eggs, from 1971 to 1995, against the parts per million (ppm) concentration of DDE. The "standard" for comparison was museum samples of cormorant eggs collected before 1947, when DDT applications were still limited to military use. The study found that in 1971 eggshell DDE concentrations reached almost 18 ppm (concentrations above 15 ppm in eggs result in complete reproductive failure).[25] Eggshells were 30 percent thinner.[26] Moving farther up the food chain, the high pesticide levels in cormorant eggs may also have posed risks for the mammalian and avian predators that fed on the contaminated embryos and resulting young chicks.

Everywhere the double-crested cormorant nested, migrated, and overwintered its numbers waned. In order to maintain a constant population level, cormorant pairs had to produce and successfully raise at least 2 chicks each breeding season; in reality their reproductive output was statistically a tenth of that figure, about 0.2 chicks per pair per year.[27] The yearly crop of new chicks was simply not large enough to replace the annual numbers of adult deaths. The double-crested cormorant was on plan for its own extinction. After the eventual banning of its use, however, concentrations of DDT (and DDE) in the environment slowly declined. As a result, eggshell thinning decreased until 1989, when DDE concentrations leveled out at 2 ppm, allowing eggshells to return to their pre-1947 thickness.[28] Cormorants' eggshells were now strong enough to protect the embryos within, but DDT still posed other problems.

A later study, conducted by the US Geological Survey and the US Fish and Wildlife Service in 1994 and 1995 on Lake Michigan in Green Bay, Wisconsin, found that DDE was continuing to be the major culprit in the cormorant's decline in nesting success. It found that even at DDE concentrations low enough *not* to cause eggshell thinning, 32 percent of the contaminated cormorant eggs failed to hatch. The study suggested that DDE contamination, above any other factors, posed the largest risk to developing embryos, almost regardless of its ppm concentration.[29] DDE was and continued to be dangerous stuff.

Probably the most heart-breaking consequences of DDT, DDE, and DDD associated with our cormorants are the terrible birth defects and deformities they cause, which are "spectacular and often grotesque."[30] They stem from chemical conflicts during the chick's development. Physiological miscues and misreads of foreign estrogenic chemicals in the egg can trigger all sorts of abnormal effects. The most common defect in cormorants is the crossed-bill defect. Common in the Great Lakes, the crossed-bill defect occurs when the upper and lower halves of the bird's bill do not meet but rather curve off, sometimes in opposite directions, making it physically impossible for the young cormorant to catch and eat fish. The bird starves to death when parental care ends simply because it cannot feed.

The crossed-bill defect is only one in a sizable listing of birth defects in cormorants associated with high concentrations of synthetic chemicals in the environment. In numerous reports scientists have found birth defects threatening large numbers of birds. Some chicks exhibited high degrees of edema, a buildup of fluids within the bird's body, which can kill the chick before it has a chance to hatch. The range of other defects included missing eyes, dwarfed limbs, clubfeet, incomplete skulls, missing or abnormal vertebrae, and internal organs growing outside the body. Combined with overly fragile eggshells, DDT flashed its lethality and struck down huge portions of cormorant populations.

It's also important to remember that DDT was not the only toxic agent drifting through the ecosystem at the time. The concept of what scientists call synergism—which occurs when two or more elements within a system interact so that their combined effect is greater than the sum of each of their individual effects—may be in play here. The results of some studies support the idea of synergistic effects among DDT, organophosphate pesticides, and polychlorinated biphenyls (PCBs) in that they may be responsible for amplified thinning of eggshells.[31] Similar reports show that the dramatic effects of DDT masked the more subtle toxic effects of PCBs, which didn't show themselves until DDT was banned in the United States in 1972.

Ten years prior to 1972, however, a marine biologist from western Pennsylvania who wrote informational brochures for the US Fish and Wildlife Service created an environmental stir of her own. In 1962, Rachel Carson's hugely popular book, *Silent Spring*, was published; advance sales on publication day totaled 40,000 with an additional 150,000 copies allocated for Book of the Month Club sales. It remained on the *New York Times* Best Seller List for thirty-one weeks. Known for her meticulous documentation and a sense for the interconnectedness of nature, Rachel Carson detailed the environmental problems associated with the indiscriminate spraying of DDT. She also questioned why so many applications of DDT had occurred without first investigating its long-term effects on the environment, wildlife, and human health. Carson never proposed the complete elimination of pesticides, but she did urge the study, regulation, and control of such powerful poisons. In her book she did, however, profess a deep fear of DDT and its cousins. "What sets the new synthetic insecticides apart is their enormous biological potency," she wrote. "They have immense power not merely to poison but to enter into the most vital processes of the body and change them in sinister and often deadly ways."[32]

Depending on the point of view being presented, *Silent Spring* was credited, or blamed, for the rise of the environmental movement. In 1997, the Natural Resources Defense Council compared Carson's book to Thomas Paine's Revolutionary War incitement to arms, *Common Sense,* and Harriet Beecher Stowe's antislavery tale, *Uncle Tom's Cabin,* both of which galvanized the nation.[33] President John Kennedy read *Silent Spring* and was so struck by the questions it raised that he instructed his Science Advisory Committee to look into Carson's assertions. On the other side of the issue, Dr. Robert White-Stevens, a biochemist and spokesman for American Cyanamid, referred to Rachel Carson as "a fanatic defender of the cult of the balance of nature." White-Stevens was also credited with a public statement claiming, "If man were to follow the teachings if Miss Carson, we would return to the Dark Ages, and the insects and diseases and vermin would once again inherit the earth."[34] That's powerful rhetoric; after all, a great deal of money was at stake. But regardless of where people stood, *Silent Spring* raised important questions that were not to be left unanswered.

A press release issued by the US Environmental Protection Agency (EPA) on December 31, 1972, began, "The general use of the pesticide DDT will no longer be legal in the United States after today."[35] William Ruckelshaus, the EPA's first agency head, defied the science advisers, who recommended a phase-out program as opposed to a direct ban shutting down all but emergency uses of DDT. It was the potential end of an era, but DDT producers did not go down without a fight. Several fights, in fact, erupted in the form of lawsuits brought before the US Court of Appeals for the District of Columbia and the Federal District Court for the Northern District of Mississippi. Suits were also brought against the ban by a number of conservation groups, such as the newly formed Environmental Defense Fund. The chemical companies wanted the ban overturned, and the conservationists claimed that the ban did not go far enough. The suits were eventually consolidated, and in 1973 the Court of Appeals ruled that the EPA was correct in its banning and deregistration of DDT.

It was now a new world, or at least a new USA. The United States was not the first or the last nation to ban the notorious insecticide. Norway and Sweden had acted to eliminate DDT use two years earlier, in 1970. The United Kingdom, however, continued to use DDT until 1984, and South Africa, home to malaria-stricken populations, waited until 1996 to ban its application. The largest international agreement recognizing the immense biological power of pesticides, the Stockholm Convention on Persistent Organic Pollutants (POPs), was signed by the United States in 2001. Signed by a total of 122 nations, the goal of the treaty was the complete global elimination of POPs, including PCBs, dioxin, and DDT. Ruckleshaus's bold administrative decision, though not the very first, fashioned the mechanism for the return of American wildlife, including the double-crested cormorant, to its past grandeur.

This "return to normal," however, would not happen overnight. In a 1979 forestry study on the effects of DDT on nontarget organisms, the author called DDT "the most widespread and pernicious of global pollutants."[36] The legal production and spraying of DDT in the United States ended on the last day of 1972, but DDT's persistent nature would allow it to show its ugly face for some time, perhaps decades. For instance, in 1971 alone, the year prior to the ban, about 13 million pounds of DDT were applied to crops in the United States.[37] These tons of poison would live in the environment for some time. Investigators looked at the half-life of DDT in the environment. This is the time period required for half the DDT to leave a particular site, either from organisms or from a matrix such as soil. In rainbow trout, for example, the half-life for the elimination of DDT is about 160 days, and in soil, particularly organically rich soil where DDT is virtually immobile, the half-life is between 12 and 15 years.[38] And that is only half the DDT, then half of that amount, and so on, for multiples of these time periods. And these figures do not take into account reintroductions of DDT to areas through surface runoff, contaminated groundwater, and transport of the pesticide through the atmosphere. Recovery appeared to be a long process, but William Ruckelshaus, in his controversial decision, and Rachel Carson in her foresight, both fought and ridiculed by arrogant chemical interests, at least turned the world in the right direction.

The decision to ban DDT was vindicated as soon as 1975, just three years later, when an EPA press release reported, "Residues of the pesticide DDT in the food supply, human tissues and in the environment have declined in recent years especially since the chemical was banned for major uses by EPA in 1972."[39] Contamination by DDT in some birds had declined but still remained at dangerous levels. As time went on, though, some figures improved dramatically. A 1997 Environmental Defense Fund press release reported that twenty-five years after DDT was banned bald eagle numbers showed a tenfold increase from

500 breeding pairs in 1962 to about 5,000 pairs in 1997, peregrine falcons from 39 pairs in 1975 to 993 pairs in 1996, and ospreys from fewer than 8,000 pairs in 1981 to more than 14,000 in 1994.[40] The new numbers again made the case for Ruckelshaus and Carson.

On the cormorant scene things began to look up as well. The Canadian Wildlife Service (CWS) published some very encouraging figures regarding the cormorant's recovery. Counts conducted during the first year after the DDT ban found only 125 breeding pairs of double-crested cormorants on the Canadian Great Lakes. By 1981 the CWS logged 907 nests on the Great Lakes in Ontario and 2,221 nests on the same sites just four years later. When new nesting sites were added to the tally, the cormorants were building their numbers at an annual rate of 39 percent. During the breeding season of 1993, 38,000 pairs of double-crested cormorants settled on the Canadian Great Lakes.[41] Cormorants were indeed back in force.

The mosquito completes its life cycle in days, has thousands of instantly free-living offspring, and produces so many generations that local genetic mutations race through the populations. Conversely, mating pairs of double-crested cormorants produce just four eggs, which, with luck and intense parental care, result in two juveniles ready for their first fall southern migration. New mutations take millennia to anchor themselves in the bird's collective genome, so, as clever and resourceful as they are, cormorants were unequipped to muster the genetic flexibility to beat DDT before it nearly exterminated the species. In this case, cormorants weren't challenged and attacked by nature, competitors, predators, or even substances that exist in nature. The species' own remarkable success as a hunter of fish, prey that became poison to its predator, worked against the cormorant and was its near-permanent downfall. Cormorants stopped reproducing. Nature had left them no more options. But since it was human ingenuity that created the toxic environment and triggered the crisis, it would take human conscience to eliminate it. The process took three decades but finally came to pass on the last day of 1972.

PART 2

*The Great Lakes:
A Place of Conflict*

6

Champlain, Native Peoples, and Henderson Harbor

WAS IT HENDERSON HARBOR area men who killed so many cormorants that 1998 summer night on Little Galloo Island? The massive attack on the island bird rookery was clearly a calculated act with the specific intention of removing cormorants from the local environmental equation. The shooting had to be associated with Henderson Harbor; a large percentage of its residents wanted the cormorants gone. Henderson had long been an important part of Lake Ontario's history and the story of the entire Great Lakes system. These men may not have had history on their minds when they pulled the triggers, but residents of the Great Lakes town had a definite history of defending what they felt belonged to them, and so Henderson was closely linked to Lake Ontario's past and future, as were the cormorants.

In his 1986 book *The Late, Great Lakes,* William Ashworth describes the Great Lakes as the "Fifth Coast," the four other North American coasts being the Atlantic, Pacific, Caribbean, and Arctic.[1] Ashworth's Fifth Coast is situated not on the edge but on the midline of the continent and sports an amazing shoreline more than ten thousand miles long. Ashworth also asserts that, because of their vastness, the Great Lakes and what they stand for have been sorely misunderstood in a number of ways.

The first of three misunderstandings was that the Great Lakes are lakes at all. Are they really lakes or inland seas? They are neither, really, but something in the middle; their huge proportions put them in a classification of their own. Ashworth prefers the term *mers douces,* or "sweet seas," a phrase coined by the famous French explorer Samuel de Champlain to describe the lakes. Jean Nicolet, another French explorer of the same period, clearly thought of the lakes as seas. In fact, in an exploratory expedition, Nicolet paddled west with a group of Native American Hurons in canoes from the Straits of Mackinac in search of the widely sought after Northwest Passage, a waterway to the Pacific Ocean and the trade and wealth of the Orient. When he reached the far shore he fully expected to step from his canoe onto the shore of China. What he had discovered would later be called Green Bay, Wisconsin, on Lake Michigan.

The second misunderstanding concerning the Fifth Coast was that the lakes were "treated as though they were infinite,"[2] as were all the natural resources associated with them—fish, game, furs, timber, and the water itself. The fur bearers, including the once ubiquitous beaver, were gone by the 1830s, the big trees sawmilled out in the "Big Cut" by 1900, and the inestimable commercial fisheries depleted by the 1950s, revealing the fishing industry's relationship to the environment and the double-crested cormorant's natural history.

Associated with the unconditional concept of the lakes being infinite in scope was the third misunderstanding: the Great Lakes were also changeless, also an unconditional concept. Some saw the huge lakes as having always existed and believed they would exist through the end of time without variation. But in geologic time the Great Lakes are new, fresh, and not yet fully evolved and developed. And it is the very idea of their vastness that guarantees that they will change: rainy seasons in some areas in some years, droughts in others, snows or no snows, poor reproductive seasons for some species in some lakes and not others. These changes cannot remain isolated forever. A series of small changes in a system will ultimately link and build together over time, resulting in major alterations, sometimes before they are noticed, as in the reported "death" of Lake Erie due to pollution in the 1960s. Nothing exists that never changes.

If we look at the system itself, the five Great Lakes, Superior, Michigan, Huron, Erie, and Ontario, form a basin of more than 200,000 square miles,[3] an area so large in scale that it obscures the mind's ability to visualize it. The Great Lakes comprise the largest freshwater reservoir in the world, aside from the polar ice caps, and with global warming effects accelerating, that value might need to be reconsidered. The statistics on the lakes are astounding. According to the EPA, the 94,000 square miles of the Great Lakes water surface contain more than 5,400 cubic miles of water, about 21 percent of the world's freshwater and more than 80 percent of the North American supply.[4]

Although the basin is a system existing as a unit, each lake has its own claim to fame. Superior is the largest and deepest, averaging 483 feet and reaching the maximum depth for the system at 1,332 feet. Erie is the smallest of the five lakes, averaging 62 feet deep, with a maximum depth of a little over 200 feet. Lake Michigan sets itself apart from the others by being the one lake in the system whose shoreline lies completely within US boundaries (the other four lakes each share portions of their shorelines and waters with Canada). Huron lays claim to the largest land drainage area, including the productive farmlands of the Saginaw River basin. Lake Ontario has about 25 percent less surface area than Erie, but because of its greater depth it holds more than three times the volume of water, an important distinction as we will see later when we characterize the conflicts between fishermen and cormorants. The variances in geography and geology across the basin produce inconsistent microclimates, resulting in differences in how people settled and used the lakes and the land around them. And as for people, in modern times about thirty-three million people live and work there, supporting 10 and 31 percent of the US and Canadian populations, respectively, and having a great influence on the Great Lakes environment.

When we look at the European settlers in the Great Lakes basin, and those who followed them, we realize how much they owe to the native people who first entered the area as far back as ten thousand years ago, around the time of the last ice age. Some of the many tribes that eventually resided in the area carried the familiar names of Chippewa, Huron, Cree, Fox, Ottawa, Sioux, and Iroquois. Other tribes, such as the Monsoni, Miami, Winnebago, Assiniboin, Conestoga, and Tuscarora, perhaps less familiar in some circles, played equal parts in settling the area. These people, who perhaps had migrated across the land bridge from Asia or had somehow managed impressive voyages across the Pacific Ocean, settled the width of the basin and established many villages along the shorelines of the lakes. The bands and tribes were sustained by hunting, cultivating crops such as corn, squash, and beans, and harvesting naturally found foods such as wild rice. But the villages could not survive and grow on these foods alone. They needed more protein, namely, in the form of fish.

To access the fish of the lakes and streams, many bands of native people navigated the local waters in finely crafted canoes. These shallow-draft boats, reaching forty feet long with a beam of five feet for their freight carriers, were constructed of a birchbark skin sewn with spruce roots over a bent white cedar frame. Tribal craftsmen caulked their seams with pine pitch, a sticky sap drawn from native pines, mixed with charcoal for waterproofing.[5]

The basin's lakes, rivers, and streams developed into the tribes' main transportation routes, and their explorations by canoe constantly opened new land for farming and hunting. The tribes regularly built their villages with easy access to lakes and rivers offering productive fishing grounds, which provided them with much-needed year-round food. From their canoes, native fishermen worked seine nets of twisted and knotted tree bark to trap and harvest the plentiful trout, sturgeon, whitefish, and other species. During the tough times of winter, with waterways frozen, the native fishermen hunted fish with spears through holes cut in the ice. The tribes also designed methods to preserve excess catches by air-drying the fish flesh on racks or leaving them to freeze in the winter cold. As the populations grew and the native people depleted the local fish, game, and crops, they moved their villages, with the life cycle of a village running about ten years.

Eventually, after European explorers returned to their homelands and told tales of the vast lands and productive waters of the Great Lakes, Europeans returned, this time as settlers rather than explorers. Many of the new arrivals traveled from nations that were historically linked to fishing. Some fish species in the Great Lakes, such as trout and sturgeon, resemble those the settlers knew from the Old World, so these fish graced the tables in their homes and became the mainstay of their diets. And then, as the European-descended settlers increased their numbers and their settlements, they looked to the local Native American tribes, with their learned fishing expertise and traditions, to supply fish for their tables. So before long fish became one of the most traded commodities on the lakes, setting the pace for future trade and industries.

Henderson and Henderson Harbor were both named for a wealthy

New York City land speculator, William Henderson. In 1795 Henderson purchased what was then called "Lot #6" from William Constable's Eleven Towns Survey or, as it was sometimes known, the Black River Tract. The price was at the remarkable, speculative rate of just one dollar per acre. Henderson then had the lot surveyed by Benjamin Wright, who described the parcel as a "pretty good town" with an excellent harbor and abundance of good timber.

The Henderson area, located on eastern Lake Ontario, played an important historic role as far back as 1615 when Samuel de Champlain, often portrayed as the Father of New France, entered the area from Canada, accompanied by a French and Huron Indian war party set on attacking the Iroquois. The party landed at Stony Point, near the future town of Henderson, where the lake passage was extremely rough going. It was a place where Indians carried their canoes around the point for safer travel, rather than risk negotiating the lake itself. Champlain and his men hid their boats before moving inland into the heart of Iroquois territory. The attack proved to be a decisive military failure in which the Iroquois warriors defending their homeland defeated the French and Huron Indian band. The group was forced to retreat from the area, bearing the wounded Champlain along with them.

Jumping ahead, about 180 years later, the earliest permanent white settler in the Henderson area, probably David Bronson, built a log cabin about 1795 in what was later to become the center of town. There he planted turnips and herdgrass but was flooded out by the lake at high water. He then moved upland and planted the first known orchard in the region. After that, beginning around 1803, the colonization of the Henderson area was bolstered by a group of robust Scotsmen from Perthshire who settled in the area. These rugged pioneers thrived in the pioneer way and cleared the land along the shore for farming .

In the course of its evolution, Henderson's name went through several versions. Henderson Bay was called the Bay of Niaoure by the French. The town center was known as Salisbury's Mills after Ludwick Salisbury, an early settler, and Henderson Harbor was once called Naples, after the picturesque city in Italy. According to various historical records, the town of Henderson was formally founded as a political entity in 1806, with the first town meeting held in March, where Jesse Hopkins was selected as the first town supervisor.

The plentiful timber that Benjamin Wright described in his survey was certainly put to good use. Besides the construction of houses and barns, between the War of 1812 and the Civil War in the 1860s, Henderson shipwrights built more than forty schooners specifically designed for the expanding shipping trade on the Great Lakes.

The War of 1812, called the Second War of Independence by some historians, played an important role in the development of Great Lakes settlements, including Henderson. In *The War of 1812: A Forgotten Conflict,* the author, Donald Hickey, an authority on the republic's early history, describes the struggle as an obscure war with a string of mysterious and varied causes. Britain's actions in taking and impounding American merchant ships and their cargoes, and the forceful impressment of US seamen into service aboard British naval ships, sometimes within sight of American shores, inflamed American emotions and threatened the honor and prestige of the new nation. The British presence was so strong in Canada that in the United States there was a constant notion of invading and conquering its northern neighbor and reducing or dissolving the close historic bonds the British maintained with American Indian tribes.

Economic matters, as they always do, played a role in the building animosity toward England. In a resumed conflict between France and Great Britain the United States upheld its neutrality, except for its backdoor trade with French interests, which created huge fortunes for American shipowners and at the same time aroused the financial jealousy of British maritime firms. As a result, the British navy seized American ships, cargoes, and seamen from American ships, many of whom were British deserters taken along with legitimate American citizens. American shipowners' losses mounted and insurance premiums soared. War talk echoed through the halls of Congress. After scores of bills, deals, legislative amendments, committee actions, and votes, the Senate passed a bill declaring war on England on June 18, 1812.

Several points of view consider the War of 1812 to have been a futile tug-of-war between British and Americans that resulted in little change and merely maintained the political status quo. The war has been summarized as an "ever-escalating game of maritime one-upsmanship" in which both sides built cargo ships and armed frigates but held them to be so valuable and agonized so much over their possible loss that the fleets rarely engaged each other directly in battle.[6] On the lakes, the fleets of both navies were instead more involved in the landing of troops, and ships maneuvered and juggled for the best position from which to support and supply their foot soldiers in battle rather than facing each other directly.

When war was finally declared, a number of Great Lakes settlers, who still carried the memories and tales of the ferocious, bloody Indian attacks of the Revolutionary War, left the area. Of the residents who stayed, in towns like Henderson, many people saw little value in fighting the British and the town voted as a political entity that it would not allow American military exercises and preparations for battle to be conducted within its jurisdiction. But in confronting the enemy face-to-face, Henderson townsmen did not hesitate to engage the British and their Indian allies. They joined existing American forces and fought in local battles, as in the Battle of Sackets Harbor, defending what they felt belonged to them, especially when their families, homes, and financial interests were threatened. At one point the women of Henderson took up arms to defend the town. When invading British and Indian forces penetrated the town while the menfolk were away, the women confronted them with muskets, driving the surprised troops and warriors from the area.[7] During the war, Henderson residents exhibited a posture of "rugged individualism," standing up for what was theirs in spite of, and in the face of, what the government proclaimed. The family lines of Henderson residents may not extend as far back as the War of 1812, but they project an attitude seen in later times, as a portion of the Henderson population in their minds defended their livelihoods against marauding cormorants and what they considered to be continued government neglect.

Towns in the eastern basin of Lake Ontario during the war against Britain set traditions that tied their economies even more closely to the waters and resources of the Great Lakes; the lakes played a vital role in moving troops, supplies, and weapons throughout the area. In the war's series of hit-and-run conflicts there was hardly a case in which naval forces did not play a major part. Sackets Harbor, a few miles northeast of Henderson, and Henderson Harbor were major sites of American boatbuilding during and after the war. The Sackets Harbor boat works built the powerful sixty-six-gun frigate *Superior* in just eighty days. The frigates *Mohawk* and *Jones* were launched from that same harbor, and the frigates *New Orleans* and *Chippewa* were started at Sackets Harbor but not completed before the war ended in 1814 with the signing of the Treaty of Ghent.[8] Boats, shipping, and boatbuilding intensified interest in the lakes and remained important economic activities on Lake Ontario through the first decades of the twentieth century.

Aside from incidents like the Battle of New Orleans, the burning of Washington, DC, and providing the inspiration for the writing of the "Star-Spangled Banner," the War of 1812, on the whole, accomplished very little. Neither side paid reparations or made concessions. The Treaty of Ghent merely reestablished the conditions that had existed prior to the war, and it was the last war fought between England and the United States.

In addition to contributing to the war effort in its own way, at various times Henderson's location in upstate New York also became the center of social, religious, and cultural change. Before the Civil War the area served as a "station stop" in the famous Underground Railroad, whose efforts were aimed at secretly moving escaped slaves from the South across the border to the safety and freedom of Canada. The 1820s–30s marked a time in New York's history of increased religious fervor and a revivalist return to evangelism, Christian scripture, and salvation issues. The period is sometimes referred to as the Second Great Awakening. The renowned revivalist preacher and educator, Charles Grandison Finney, conducted popular evangelical meetings in the Henderson area. Numerous churches and societies, such as the Washington Lodge, part of the Freemasonry network, the Second Baptist Church, the Henderson Universalist Society, and the Swedenborgian Society, followers of Swedish scientist, inventor, and philosopher, Emanuel Swedenborg, each established memberships in or near Henderson. The convergence and acceptance of so many philosophies, religions, and opposing points of view no doubt added to Henderson's atmosphere of independence and free thinking.

After the Civil War, and particularly after World War II, Henderson residents and businessmen turned their dependence on Lake Ontario from commercial shipping to recreation. The Industrial Revolution, with its increased exploitation of machinery instead of labor-intensive toil, helped build America's middle class, the forty-hour work week, and the concept of leisure time. Add to that the production of reliable, reasonably priced automobiles and the construction of passable roads and the way was cleared for the building of summer houses along the shores of Lake Ontario and the expansion of recreational opportunities. Leisure time and recreation were translated into sportfishing, a pastime that grew in direct proportion to the increased popularity of local towns and hotels as vacation destinations. Lake Ontario also became nationally known for its famous guides, who would row or sail patrons, including wealthy and at times international public figures, out to nearby productive fishing grounds. The Galloo Islands group was one such destination, bringing the hardworking, nesting, double-crested cormorants into the picture.

It was then, in the 1950s–70s that the number of double-crested cormorants crashed. The wide use of DDT as an agricultural insecticide decimated the fish-eating bird population. The ingestion and metabolic breakdown of DDT weakened their eggshells and caused widespread, catastrophic reproductive failure. In the 1980s, the decade after DDT was finally banned, the birds came back in force at essentially the same time that the recreational fishing industry declined. Unfortunately for the cormorants, the game fish disappeared in a *perceived* proportion directly related to the boom in cormorant colonies. Recreational fishermen, guides, and other related interests chose to ignore the hidden, less obvious effects of overfishing, industrial and agricultural pollution, the introduction of nonnative competing fish species, and the ruin of historic fish-spawning runs and regularly ignored alternatives presented by conservationists and environmentalists.

Regardless of the fishing industry's destruction of its own future, cormorants were again targeted as the ultimate evil on the Great Lakes.

7

Fishing America's "Fifth Coast"

THE DODGY CONFLICTS in which double-crested cormorants find themselves involved time after time revolve around food, and food to them means fish. Although cormorants sit at the top of their food chain as aquatic predators, adult trout and salmon and some other fish do the same job. As top predators along with man, these fish and the double-crested cormorants face a tangle of obstacles and changes that complicate the overall food chain in ways scientists and fishermen may never fully understand. But the common factor in this jumbled food chain scenario is the fish.

Humans, of course, always had food options other than fish, but as in the case of early European settlers, the plentiful fish stocks often became the mainstay of their diets. Along the lakes and their tributaries, humans always had the options of raising crops and tending livestock and frequently lived directly off the wild bounties of the land; they could live without fish. The cormorants, salmon, and trout had no other choices. Their environmental niche was in or on the lakes, feeding on fish, and the only other alternative was their extinction. In a significant sense, then, since no species exists in a vacuum, the story of the cormorant's history on the lakes is the history of the fish and the fishermen. Without understanding the role of sportfishing and its big brother, the commercial fishery, it is impractical to attempt to understand the cormorant's second coming and present niche on the lakes.

The birth of what Samuel de Champlain referred to the "sweet seas," *mers douce,* the Great Lakes, is a relatively recent occurrence on the geological calendar. The period of ten to twelve thousand years ago corresponds to something like the time in the past when you began reading this sentence. It is essentially *now*. And the extreme infancy of the Great Lakes shows itself in the very few kinds of fish found in them. The vast expanse of more than 5,400 cubic miles of freshwater boasts only 136 to 150 different fish species, depending on the data source and classification scheme. In the dozen or so millennia that mark the history of the lakes only 150 species at most have had enough time to call the Great Lakes home. Even accepting the difficulty of comparing other water systems to the largest freshwater system on Earth, many other freshwater basins support far more fish species than North America's inland seas. According to the website Fishbase.org, an online relational database of fish species developed through collaboration with the United Nations, the European Commission, and nine research institutions, our own Mississippi River basin claims 226 fish species. South America's massive Amazon sustains 1,214 and the Orinoco River basin 540 distinct species of fish. The site also lists Vietnam's Mekong River as having 785 species. And again, the vast waters of the Great Lakes come in with only 150. In the comparisons, if we allow an adjustment for a greater diversity of life occurring in tropical climates, 150 species is still a meager showing.

Numerous studies of the ecosystems of the Mississippi, Amazon, Orinoco, and Mekong reveal how fragile their food chains are, and with as much as an order of magnitude of difference in the number of species, it's easy to imagine how tenuous the relationships are among the relatively few species of the Great Lakes. And it is just as simple to imagine how easy it would be to upset their still-evolving natural balance. And what could upend that balance more than humans removing so much high-quality protein from the ecosystem in the form of predators like trout and salmon. And all that bounty, unregulated, and free for the taking?

Throughout the literature on the early exploration of the lakes, North America's Fifth Coast, writers recorded an almost unbelievable abundance of fish. The numbers were uncountable. And the sizes! The extent of the fish populations was unlike anything Europeans had seen in their homelands. Tons and tons of forage-sized fish and huge prowling monsters such as the primitive-looking, long-lived lake sturgeon, measuring nine feet long and weighing-in at close to four hundred pounds. It takes years of growth for sturgeon to achieve such proportions, the price for their size being their slow progress to sexual maturity. Males live to be 55 and mature in 8 to 12 years; females live for 80 to 150 years but do not mature until they are 20 to 25 years old. Their slow reproduction rate made the lake sturgeon susceptible to elimination later through overfishing. Along with sturgeon, Europeans found the burbot, or ling, a freshwater version of the saltwater cod; the blue pike, also called sauger, a type of perch; the cisco, a foot-long fish also referred to as lake herring; and the ubiquitous and tasty yellow perch. And there were others yet.

Among the others were the three piscivorous species, fish eaters, the ever-important predators: Atlantic salmon (in Lake Ontario), lake trout, and lake whitefish, another type of trout. The three predators, harvested from pristine Great Lakes waters, were always in great demand for the dinner table by those who lived along their shores. These fish were of great importance to the native peoples, later to European settlers, and the lake whitefish has been extremely popular in more recent times.

In her book *Fishing the Great Lakes: An Environmental* History, *1783–1933,* Margaret Bogue relates the tale of the Great Lakes Atlantic salmon.[1] This salmon, sometimes called the Lake Ontario salmon, was a landlocked form (subspecies) of the true Atlantic salmon and, restricted in its range by Niagara Falls, was found only in Lake Ontario, while the other two predators were found in all five lakes. Known for its huge, dense, upstream spawning runs in the many creeks and rivers feeding Lake Ontario, this valuable salmon was harvested by any means possible. Along creek shores fishermen used clubs, shovels, pitchforks, spears, and bare hands, as well as traditional fishing gear. Thousands and thousands of pounds of salmon were captured and salted during the spawning runs and then packed into hundreds of barrels for shipment to other settlements and towns.

Bogue discusses the Atlantic salmon's outcome as an operative model for the fate of many other fish in the Great Lakes system. The Atlantic salmon, it turns out, was a fish of firsts. It was the *first* of the commercially valuable fish to be "fished out," meaning its numbers were so low that the Atlantic salmon was no longer a profitable species to pursue. Then, in an overdue attempt to preserve the Atlantic salmon fishery, it was the *first* to be the target of regulation by both Canadian and American (primarily New York State) lawmakers in the nineteenth century. Regulations included the establishment of seasons and harvest limits to protect the spawning runs. For the most part, though, regulations were ignored by fishermen or went unenforced by officials. The last, and saddest, of the Atlantic salmon's *firsts* was that it was the first documented extinction on the Great Lakes. So documented and so complete was its extirpation from the lakes by the 1840s that Bogue classified it as a paradigm for the decimation of a fish species. Much like the stages of grief used by psychologists, Bogue developed six elements of extinction for many other Great Lakes fish. They can be summarized as follows.

1. Changes in land and waterway use such as logging and damming throughout the drainage basin
2. Powerful market demands, which bring strong harvest pressures to bear on a single species
3. The development of more efficient fishing techniques and gear aimed at taking a particular fish species
4. The lack of effective regulations and protective, enforceable controls for a pressured species
5. The lack of scientific studies providing data and conclusions for use in drafting practical legislation, seasons, and limits
6. The setting of administrative and political priorities, which always place the protection of particular fish stocks low on the list of priorities[2]

It was not by design but rather through apathy, neglect, and greed that these elements did, in time, fashion the demise of other fish in the lakes. The first element alone, changes in land and waterway use, was a force formidable enough to bring about enormous changes. Just as it occurred in New York State's Catskill Mountains, the removal of trees along riverbanks eliminated shade from the intense summer sun and raised water temperatures above the tolerance level of many aquatic insects and the juvenile fish that fed on them, killing off both. And as thousands of logs were rafted and floated downstream to the lumber mills, the cut timber gouged away the vital gravel beds where native trout and salmon had spawned and in which their eggs and fry had developed for the past hundred centuries. New generations of fish were destroyed before they left their birthplaces, with little hope of being replenished in years to come.

William Ashworth, in his environmental history of the Great Lakes, described its problem as one of human perception: "We always treat things as we perceive them; and if they are perceived wrongly, they will be treated wrongly as well."[3] Ashworth describes a simple sentiment and paints a compelling picture. While he was writing his book, in the late 1980s, Ashworth toured the shores of the Great Lakes. He saw how people and industries had defiled them. It was a gloomy tale of one "squalid grey waterfront" after another. The name of the city didn't matter—Cleveland, Toledo, Duluth, or Buffalo—the shorelines, some of them former bathing beaches, were crusted with fifty-five-gallon chemical drums, concrete blocks, and rusted steel beams and had been left with rusting wrecking yards, abandoned warehouses, oily creeks, and tons upon tons of garbage. Who let this happen to *their* lakes? And the complex question arises: when did this negligent, arrogant, everything free for the taking, cavalier attitude toward the lakes and their resources, including our double-crested cormorants, come about?

The record of fishing along the Great Lakes goes back as far as five thousand years ago. Archaeologists have found evidence of Indian cultures existing in the Great Lakes basin dated to 3000 to 2000 BCE, long before any contact with European explorers and settlers took place. In the artifacts, researchers have found spears, gaffs, hooks, and traps, tools designed specifically for harvesting local fish.[4] These precontact fishing activities obviously supplied food for the villages in addition to trade goods for the native people's essential barter networks. Some scientists maintain that fishing helped define individual Native American cultures, as well as the tribes' settlements and their patterns and distributions.

These early people frequently used the lakes and streams to the limits of their practical capacities and at times temporarily drained the resources to the point where the tribe or band was obliged to move on. The rudimentary and primal way of life practiced by early Native Americans by its very nature set limits on how many individual band or tribal members a certain area could support; the resources were not unbounded. But regardless of their needs, the tribes respected the lakes, their waters, and the wildlife. The natives of prehistory were most likely the first conservationists and caretakers of the lakes. Their animistic beliefs taught that animals, as well as man, had souls and that even food animals destined to die at the hands of a hunter or fisherman should be respected. Ritualistic thanks and offerings had to be given to powerful spirits, like the Owners of Nature, for providing the game and fish they took as food. In time the native peoples' beliefs would clash with those of Europeans, who believed that only man has a soul and that the seemingly limitless fish and game belonged to man alone, as if they were free for the taking.

Commercial fishing on the lakes appeared about 1820, after which the industry expanded and built on itself by about 20 percent per year. The lakes were still so productive that, even outside the circle of commercial fisheries, farmers, trappers, and tradesmen continued to fish to further support their families. The fish were so thick and the harvesting so productive that very little fishing equipment was needed. A boat and a net or spear worked by one man could produce harvests large enough to feed a family and set aside a surplus to be sold or traded to neighbors.

The launch and early expansion of commercial fishing were boosted in large part by the growth of transportation systems, specifically the canal building of the early 1800s. The opening of the Erie Canal in 1825 provided a conduit through which fresh and salted fish could reach large, hungry eastern markets, such as New York City, by way of the Hudson River. Shortly after the Erie Canal began operation, the American Fur Company established the first organized commercial fishing venture on Lake Superior in the 1830s. Managed by European and Scandinavian immigrants, the company used lengths of gill nets to capture all manner of fish, their primary target being the lake whitefish, cousin to a European favorite. The nets and harvests were handled for the most part by Native Americans employed by the company, who had firsthand knowledge of the lakes and the fish and had been displaced from their traditional tribal hunting and fishing grounds by the felling of the basin's forests by lumber interests.

The American Fur Company eventually failed, but other businesses, like its competitor, the Hudson Bay Company, continued on, with new fishing undertakings springing up to exploit the free, unregulated resource in all of the five lakes. From its inception, profits from commercial fishing efforts put more cash into people's pockets; enough cash to buy more goods, pay down existing debts, and in some cases enough income for even part-time fishermen to buy or create farms and businesses of their own. Commercial fishing became an important tool for creating wealth, fortune, and capital on the Great Lakes.

As human populations grew with the new national prosperity of the mid-1800s, the market for fish grew with it. In the distribution network that evolved to serve the growing industry, the city of Detroit became its "hub," the single largest supplier of Great Lakes fish to the country. Dealers in Detroit controlled the industry's finances, the prices paid for fish at different levels of the distribution network, and the buying and renting of fishing equipment and new technologies.

The second half of that century saw a burst of interest in new, less labor-intensive fishing methods and more efficient commercial fishing gear. As fish populations in the near-shore areas grew smaller and smaller due to fishing pressures, canoes, rowboats, and spears became less and less useful and clearly less productive. Specially designed sailboats, the Mackinaw boats, and their nets replaced the simpler gear. Then, as fish numbers thinned out even more, larger nets were needed to maintain catch levels and profits, which in turn required more hands onboard the boats.

Detroit's answer to its growing labor costs was steam. The introduction of steam-powered fishing boats saved the day—and the industry. The larger, more powerful boats could travel safely farther offshore and tow larger nets for longer periods. Steam power technology not only propelled the boats but produced onboard equipment such as power lifters, machines designed to set and recover the larger, much heavier gill nets used in deeper offshore waters. Thus, fewer men were required to harvest more fish in a shorter period of time. Fleets of gill net steamers increased dramatically. On Lake Michigan, for instance, the number of boats grew from just five in 1873 to eighty-two in 1885 and then shrunk soon after, when owners moved their boats to more productive waters.[5]

As large as the boats and nets became, the fishing industry could not keep pace with the fact that food fish were becoming harder to find. Besides being scarcer, the fish were smaller and drew fewer dollars per pound on the market. Commercial fishermen and dealers hid the scarcity factor from consumers and regulators and "avoided the consequences of overfishing by abandoning one species of fish in favor of another."[6] The fishing fleets refitted for a new commercial species about every ten years. The reporting of total poundage caught did remain stable year after year, but the profile of fish species varied, and the profits fell as less popular, relatively low-value fish now filled the nets and markets. Ashworth described the ominous conditions of the time: "The desirable species were being picked off one by one; the delicately tuned proportional relationships among species worked out by the slow turning of natural process over the last 10,000 years were coming unglued."[7] Ecological disasters were following economic disasters.

In a 2009 *New York Times* essay, writer Aaron Hirsh dealt with the similar present-day setbacks in the depleted and overworked fisheries in the Gulf of California.[8] He discussed what economists now call the "tragedy of the commons," a bioethical concept developed by Garrett Hardin, a controversial ecologist who died in 2003. In an ideal economy the "sole owner" of a fishery "will manage it to maximize its total value over time," leaving a lot of fish in the water to breed and harvest at a later time. Hirsh likened the philosophy to an individual who would "withdraw the biological equivalent of interest, without reducing the capital." The "tragedy" comes into play when zealous competition among multiple, independent "owners," or fishermen, creates a fear that one will take everything today leaving nothing for the others tomorrow. Therefore, each fisherman takes every fish available today, leaving little for anyone tomorrow. Hirsh's discussion of fishery economy theory is a perfect description of the Great Lakes fisheries of the nineteenth century.

Steam power and new fishing technologies hauled in tons and tons of fish flesh, not all of which was considered marketable table fare at the time. Lake sturgeon, for example, which had sustained the native peoples for centuries, were treated as what is now called "bycatch," trash fish. Prior to the mid-1860s, tons of "worthless" sturgeon were dumped on beaches and set afire like so much garbage. Fishing steamers sometimes used the plentiful sturgeon, a naturally very oily fish, as fuel to fire up their boilers, building pressure for their steam engines. After all, the fish were there and free for the taking. In 1865, a limited demand, much of it from overseas markets, developed for smoked lake sturgeon, its oil, and its rich caviar. By 1899, due to uncontrolled growth in the market decimating the population, sturgeon production dropped to about one-fifth of what it was merely two decades earlier.[9]

Another significant upshot of modern steam technology was, of course, the railroad. The 1860s was a time when the progression of the railroad and track laying expanded into the Great Lakes basin. Railroads opened markets where the new American working class had settled. Cities such as Chicago, Buffalo, Cleveland, and Rochester in the United States and Toronto and Hamilton in Canada became thriving markets for Great Lakes fish, particularly lake whitefish, due the culinary affinity recent European immigrants had for it.

Closely associated with the railroad in promoting Great Lakes fish across the country was an 1869 patent for the process of quick freezing fish in metal pans for better preservation. When later coupled with the invention of practical refrigeration and insulated refrigerated railroad cars, the nation's taste for relatively inexpensive fresh, salted, and frozen fish was nearly insatiable. In 1879 more than a million pounds of lake trout and two million pounds of whitefish were harvested from Lake Ontario alone. The greatest commercial fishing harvests on the Great Lakes were recorded in two years, 1889 and 1899, at about 67,000 metric tons, or 147 million pounds.[10] It was the zenith of the industry, unable to be sustained or realized ever again.

After reaching that pinnacle of success, Great Lakes fishermen and dealers soon needed to work harder and longer to maintain their profits. Everyone in the industry knew the stocks were being driven out of existence, but there was still great resistance to any proposed regulations and controls. Garrett Hardin's tragedy of the commons ruled the lakes.

The final blow to commercial fishing on the lakes was delivered by two invaders: the parasitic sea lamprey and the highly reproductive alewife. Probably entering the rest of lakes from Lake Ontario through the new canals, both thrived. By the 1920s–30s the parasitic lampreys attached their sucking jaws to large numbers of commercially valuable fish while alewives outcompeted many native fish stocks for available food sources. By the 1950s, overfishing, pollution, and the invaders finally destroyed commercial fishing on the Great Lakes. But despite the fact that Cleveland's thoroughly polluted Cuyahoga River actually caught fire numerous times and Lake Erie, overwhelmed by excessive nutrients and choked with oxygen-depleting algae, was declared officially "dead," the lakes were to be granted a second life.

The Great Lakes Atlantic salmon species was by any real definition extinct. The lake trout fishery had been drastically reduced, and the remaining stock, netted with parasitic sea lampreys boring into their flanks, was largely unmarketable. Delectable whitefish, another lamprey victim, fished out and diminutive in size, also retained little of their former market value.

Little hope existed for any of the fisheries while millions of hungry sea lampreys prowled the lakes. The same higher temperatures in streams robbed of their shade by loggers and no longer hospitable to spawning trout became ideal reproductive havens for lampreys. Adult lampreys swam upstream into the warming water, spawned, and died. The fertilized lamprey eggs hatched in the crevices of what was left of the warm gravel beds and produced larvae that fed on plant and animal debris for three to six years as they transformed into the parasitic adult forms. The hungry adults then swam downstream to the lake to attach themselves to the few remaining salmon, trout, lake whitefish, and others for another year or so, at the end of which the cycle repeated itself. With near-perfect breeding conditions, no natural predators, and abundant food sources, the invading sea lamprey more than thrived.

Attempts to trap adult spawning lampreys had limited success due to the large number of streams and rivers available to them. But a lampricide, a poison specific to lamprey larvae, called TFM was discovered in the late 1950s. Applied to lamprey-spawning streams, TFM kills about 90 percent of the larvae without harming other fish or aquatic life (a similar claim once attached to DDT). Applications of TFM eventually produced a 90 percent overall reduction in the Great Lakes lamprey population, which allowed the rebuilding of some fisheries, such as that of the lake whitefish, and the establishment of new fisheries, namely Lake Ontario's Pacific salmon population, introduced in 1969 and replenished annually with hatchery stock. Coupled with pollution abatement, sea lamprey control helped restore and foster a four-billion-dollar industry employing thousands of people in the Great Lakes states and Canadian provinces.

The recovery of the sport and commercial fisheries is a story in itself, and provided the impetus for much of the anticormorant rhetoric and vigilantism, so understanding the recovery phenomenon helps put the cormorant controversy in perspective.

In spite of once decreasing yields, states such as Michigan support a large number of commercial fishing entities, and, through existing fishing-rights treaties, about 150 Native American tribe-licensed fishing operations also ply Michigan's waters. Another 45 state-licensed concerns add to the mix. The tribal operations are productive enterprises, taking about 50 percent of the total commercial catch. The principal targeted species today is the lake whitefish, typically amounting to eight to nine million pounds per year but with highly fluctuating wholesale prices. Whitefish still enjoys local popularity but faces recognition and acceptance problems in most national markets, so the fishery faces many challenges as production costs continue to climb.[11]

The repeal of the death penalty for the Great Lakes, control of the sea lamprey population, and growth of the Pacific salmon fishery buoyed by hatchery stock gave birth to the Great Lakes sportfishing industry we know today. The two transplanted Pacific species discussed below, along with populations of feisty smallmouth bass and table-fare yellow perch, attracted and continue to attract anglers and angler dollars to the lakes and Henderson Harbor, nestled close to the cormorant rookery, Little Galloo Island.

Surveys show that Great Lakes states have about one-third, 3.7 million, of the country's total registered boats and that half of all Great Lakes fishing is done from these boats, including privately owned and captained charter boats. Anglers include youngsters fishing with their favorite cartoon character outfits through experienced seniors using far more sophisticated tackle and digital fish-finding equipment. In 1996, the recovering sportfishing industry boasted more than 11 million anglers sixteen years and older who fished Great Lakes and other, inland waters.

As one example, Lake Erie sportfishing came back to life after its status as a "dead" lake was turned around through regulation of commercial and municipal nutrient-rich effluents in the 1970s. The lake began producing natural plant growths as the huge oxygen-depriving algae blooms died off. Enforced controls were also brought to bear on commercial operations for harvesting Lake Erie walleye (a popular and tasty sport fish in the Northeast and Midwest), allowing that population to grow large enough to satisfy eager recreational fishermen.

The two Pacific salmon species, the coho and chinook, first introduced into Lake Michigan and later into Lake Ontario, put on muscular pounds as individuals feasted on the millions of protein-rich alewives left to expand their numbers after the predatory Atlantic salmon and lake trout populations crashed. The problem was that the two Pacific salmon species do not reproduce well on their own in the lakes and require annual restocking from hatcheries. Great Lakes charter fishing fleets grew quickly as tales of trophy salmon circulated throughout the sportfishing community. But to maintain the momentum of the fishery, millions of Pacific salmon fingerling and larger "stockers" had to be released into the lakes every year in hopes of producing new generations of angler-pleasing tackle busters. Early stocking practices produced some of the most obvious man versus cormorant conflicts, which occurred when releases close to shore attracted flocks of double-crested cormorants, which considered these easy, shallow-water offerings a perfect addition to their diets.

In spite of small stumbling blocks, the fishery grew. And since charter boats require services and supplies, numerous tackle shops, bait and ice suppliers, marinas, fueling piers, and repair docks cropped up to service the fleets. Each new business meant additional jobs. Fishing boats, in addition to captains, required crews and first mates, repair services needed mechanics, and the other businesses hired people to pump gas and diesel, prepare bait, and handle ice. So naturally each boat and on-shore business had a payroll to meet. The entire economy was fueled by anglers who laid down their dollars to catch their legal limits of trophy Pacific salmon, walleye, yellow perch, and smallmouth bass. These same anglers slept in local hotels and motels, dined in the restaurants, and gathered in the taverns. Most of these family-owned and operated enterprises would never have existed in the first place and would not continue to exist without easy access to Great Lakes game fish. Heaven help anything or anybody that got in the way.

A 2002 survey of Michigan's Great Lakes charter fishing industry conducted by the Great Lakes Sea Grant Network reveals the scale of the area's sportfishing industry.[12] The survey recorded 468 licensed Michigan captains. Of the 242 respondents, 95 percent owned their boats (as opposed to leasing or being employed as its captain), with captains typically having fourteen years of licensed charter-fishing experience. In spite of their experience, because Great Lakes chartering is a seasonal business, for about half the responding captains, charter fishing was an important secondary rather than a primary source of income.

But the charter boat industry wasn't always about the money. Survey results also showed that a vast majority of the captains ran a charter service because they liked the work and enjoyed helping people learn to fish. Afloat, for the average 2002 half-day fee of $338, skippers provided their clients with bait, tackle, ice, and the cleaning of their catch. Other services such as photos, videos, food, and lodging were available from some charter businesses at additional cost.

Charter fishing was not an inexpensive business to run. The average price of a boat fitted out for chartering was about $82,000 and a number of businesses owned more than one boat. The average annual income in 2002, after expenses, for owners who faced no boat-loan payments, was $4,814 per boat. As a group, the 468 Michigan charter captains in the same year generated just over $10 million in gross sales.

The charter boat owners had a lot invested in sportfishing. Virtually every dollar of income in many lakeshore towns was in some way derived from the fish and the anglers who came to catch them. When the fishery began to decline and the number of visiting fishermen with it, the captains and other business owners looked for an answer. In their minds it wasn't the fleets of boats carrying up to six clients each taking their legal bag limits day after day. It wasn't increased pollution and runoff contaminants from the towns and businesses. And it wasn't stream and river areas damaged by shore development. It was those damned cormorants eating *their* fish.

Anglers and the media often argue that cormorants are not native to the region and are exotics, with rumors circulating that cormorant populations were deliberately unleashed by the Japanese.[13] Where exactly were the double-crested cormorants when all this was taking place? Their distant history on the Great Lakes is a little vague, but there is evidence showing that some cormorants nested in scattered colonies throughout the Great Lakes before 1900 and solid documentation that populations expanded into or recolonized many sites between 1950 and the present. Various summarized studies indicate that the tally of cormorant nests on the Great Lakes (each representing two breeding adults) ranged from 900 in 1950 to a low of 135 in 1972 and then surged to 89,000 in 1997 and 113,000 in 2005.[14] Some of the "hottest" sportfishing on Lake Ontario occurred in the early to mid-1990s, even as cormorant numbers doubled. The fish were there in big numbers, the fishermen were there in big numbers, and the birds were there, also in big numbers—in good times and in bad. The anglers took their share of the fish and the double-crested cormorants took their fair share of the fish, but the birds didn't fill their coolers and freezers with sport fish; they just ate what they needed at the time.

The captains' perception that the birds were stealing what rightfully belonged to them, perhaps a Henderson tradition carried in the air from the last war with Britain, was the driving force for what would happen on Little Galloo in the summer of 1998.

8

Ashmole's Halo

*A Righteous Model of
What Should Have Happened*

THE SPRING SUN warmed Little Galloo Island's fifty or so rocky acres and the waters of Lake Ontario surrounding it. The first cormorants to arrive that season settled on the water and the island's stony beaches at the finish of their thousand-mile migration from the Gulf Coast states and the Carolinas. The more mature and the well-traveled adults had left their southern wintering grounds in late April, while the very small number of young birds hatched the previous spring waited a few weeks and then headed north in early May.

Fewer birds gathered on Little Galloo Island this season than had appeared the previous year. This year's numbers could easily be counted in scores or perhaps mere dozens. For years now, males had acted out their display rituals to attract females, they paired off, males collected nesting materials, and females built and cemented the nest together. The pairs mated, and females laid their clutches of blue eggs in the flat stick nests. Every encounter and every action constituted a measured stage in an unchanged ritual. The ritual remained the same, but in these same years far too many of the fragile, membranous, thin-shelled eggs were crushed under the weight of a parent attempting to warm and protect the embryos. And among the hatchlings, many suffered debilitating birth defects that severely lessened their chances for survival.

Many perished of starvation within the span of only a few days because their deformities would not allow them to take food from the parents. Thirty years earlier, when the colony thrived, the chicks in two or three of the four or five eggs in each brood lived through their fledgling stage and migration to return the next year. In recent years, mating pairs sometimes produced only one successful offspring. And many nests failed completely. The youngsters that did survive the grueling round-trip migration would not reach sexual maturity for at least another two years, keeping them out of the current breeding pool. This year fewer breeding adults would form fewer pairs because they would age out and die as their natural mortality rate far exceeded their current birth rate. The double-crested cormorant was about to pass out of existence.

Despite their grimly shrinking numbers, there was a slim glimmer of hope for the birds. The year was 1973, and this spring was the first breeding season in three decades in which cormorants and other wildlife would not face the burden of additional tons of DDT entering the Great Lakes. In the seasons to come the forage fish and smaller game fish that sustained the Little Galloo Island colony would become less and less toxic. Every fish the cormorants consumed would no longer bring them closer to extinction. Each season every new generation of fish would be spawned in cleaner, purer water. These cormorants would ingest fewer of the DDT-derived DDE molecules that mimicked their own hormones and inflicted such massive damage on their reproductive cycle. It would again be possible for the cormorants' births to exceed their mortality rate. They may have edged that very short, fateful distance back from the brink of extinction—the distance that would make a difference.

Cormorants had been more susceptible to the toxic effects of DDT than many other animals on the Great Lakes, but they also recovered from the insecticide more quickly than many other species. Avian predators such as bald eagles, vultures, and various gull species, which fed on cormorant eggs and chicks, were slower to recover. So, at least for the immediate future, cormorants had a slight numbers advantage over these predators, letting more of the newly emerged cormorants survive the season on Little Galloo and elsewhere on the Great Lakes. Cormorants were facing a reprieve and, feasibly, better times, but once again their own success was about to drag them into ecological conflict.

Controversies and conflicts between double-crested cormorants and humans always involve trade and industry issues, and trade and industry issues always mean money. It's known that the everyday activities of cormorants sometimes get in the way of people making a living. The largest dispute takes place when cormorants feed, creating the perception that they are consuming the same fish as humans. Angry commercial and sportfishermen witness double-cresteds feeding individually and in large flocks and consequently blame the birds for depleting economically valuable fish stocks. It's then that the shotguns come out—blazing. It's also then that we raise the question of how the cormorant became such an enemy instead of simply an interesting bird that eats fish.

In the post-DDT era, populations of cormorants grew at phenomenal rates, locally doubling their numbers every few years. The best recovery examples were seen on the Great Lakes in both Canadian and US waters. It turned out that every fall many more cormorants departed the lakes than had arrived the previous spring. The blue shells of cormorant eggs gradually thickened as DDT-saturated fish protein faded from the ecosystem and embryos again survived incubation by their attentive parents. Cormorants were reproducing once more. They were everywhere on the lakes, and conflicts between fishermen and cormorants increased in direct proportion to the mushrooming cormorant population. And the question arose in some quarters as to why there appeared to be no end to the cormorant's steep growth curve. Alarms went off in the marinas, the taverns, the restaurants, and among the captains. Would it ever stop?

The answers were to be found through the investigation of predator-prey relationships and studies of how populations behave. Scientists look at populations and their growth from several angles. It's known that populations do have inertias of their own and under consistent conditions their sizes might continue at a constant level until an outside force acts on them. But in natural settings there are always forces acting on populations; everything is in flux. Forces come in the form of destructive storms, disease, fires, floods, droughts, new predators, and human development, all of which take their toll on reproductive potentials. These and other natural and manmade influences produce in cormorants, as in all species, chains of successful and failed breeding seasons, which keep predator-prey ratios, breeding site availability, and food supplies in balance. It's when these ratios and balances shift that drastic population reactions occur. A specific dramatic example of this reaction appeared within the cormorant populations on the Great Lakes.

One scientist who dealt with the mechanics and dynamics of population numbers was Vero Wynne-Edwards, a British zoologist and field naturalist who worked in arctic environments. Among his other honors, he was elected a Fellow of the Royal Society and Commander of the Order of the British Empire. His 1962 book, *Animal Dispersion in Relation to Social Behavior,* turned the world of evolutionary thought and theory on its ear. As a scientist, in some circles he was respected and considered to be a methodical scholar. Known also to the general public, a précis of his widely read book published in the popular magazine *Scientific American* sold 350,000 copies. He was a serious man not to be taken lightly. *Animal Dispersion* was one of the most controversial biology volumes of the 1960s and 1970s.

Wynne-Edwards faced down traditional evolutionary theorists and their strict Darwinian models of inheritance and natural selection. They saw natural selection pressures acting only on individuals and favoring some individual genetic variations over others, resulting in the rise of behavioral patterns that separate breeding populations and eventually generate new species, hence Darwin's book title, *On the Origin of Species.* But Wynne-Edwards had a different vision. Building on his studies of seabirds, he proposed that natural selection acted not just on individuals of a group or species but on the group as a unit. His concept of group selection centered on the process of how nature limited animal numbers. He argued that animal populations evolved self-regulating, programmed mechanisms such as hormonal signals, social displays, social hierarchies, territoriality, and colonial nesting to keep their numbers in check and not overtax available natural resources. Wynne-Edwards concluded that groups that were the most successful at self-regulation produced the most successful progeny and enjoyed the strongest possibilities for advancement on the evolutionary tree. In essence, he advocated survival of the fittest group rather than survival of the fittest individual.

Traditional evolutionary biologists wholeheartedly disputed Wynne-Edwards's thesis. They claimed that population sizes were affected by natural forces limiting their numbers and argued that there was no detectable evidence, either physiological or behavioral, that populations had evolved self-regulating mechanisms in which the group acted as a whole in its own best interest.

One critic of Wynne-Edwards and his group selection theory was N. Philip Ashmole, a biologist who conducted extensive studies on tropical seabirds, primarily terns, and their oceanic island ecologies. In 1961, on an expedition sponsored by the British Ornithologist's Union, Ashmole formulated the idea that heavy feeding by large breeding colonies of fish-eating seabirds could exhaust food supplies closest to the island colony, forcing adults to roam farther in their search for food. The prey-depleted zone around such islands was later termed "Ashmole's Halo" by other scientists.[1] In a now classic 1963 paper, Ashmole summarily rejected Wynne-Edwards's hypothesis of mysterious self-regulation of the production of young as "unproven and improbable."[2] Ashmole instead postulated that food and its availability were the only natural factors or forces to consistently limit the population of seabirds in a colony. He maintained that as the halo of depleted food expanded as competition for food increased, fewer and fewer adults would succeed in raising their chicks, bringing births more in line with the mortality rate, thus providing a natural check on possible ever-expanding bird populations. He also concluded that reductions in immediate food supplies would delay the onset of sexual maturity in first-time breeders, a process he referred to as "deferred maturity," which would further lessen the rate of population growth by reducing the number of recruits into the breeding pool.

Later studies of island bird populations upheld Ashmole's halo hypothesis, developed direct evidence of prey depletion by seabirds, and reinforced the idea that food supplies at breeding sites were capable of limiting population numbers.[3] These and other follow-up accounts supported Ashmole's halo concept and caused Wynne-Edwards's fashionable group selection and self-regulation ideas to fall out of favor. Since that time, Philip Ashmole's concept has been broadly accepted and has frequently been tied to environmental situations outside tropical areas, though in an amended and modified form. The revisions came about because Ashmole's original high-latitude studies did not include forces such as predation, disease, and breeding site availability because he did not encounter these factors in the tropics; they have since been folded into the concept package, which still follows Ashmole's initial model: food supplies at the breeding site control the population.

Reflecting on this evidence raises a few questions. "What happened on Little Galloo Island?" "Why didn't the cormorant colony react as Ashmole's model had predicted?" "What was it that created the situation in which the fishermen felt they were forced to kill so many birds to protect their living on the lakes?"

The first day of 1973 opened new horizons for the world's wildlife, particularly in the United States and particularly on the Great Lakes when the manufacture and general use of DDT were banned, so double-crested cormorants and other birds would face no additional introduction of the toxin into their environment and food chain. In the nearly two decades from 1973 to 1991, a span of only eighteen years, cormorant populations multiplied by a factor of three hundred, with more than eighty new colonies springing up across the Great Lakes.[4] Cormorants were reproducing at an average annual rate of nearly 30 percent.[5] Given Ashmole's generally accepted theory of feeding halos, over time decreasing food supplies should have begun to draw down cormorant numbers at breeding sites like Little Galloo Island. But the reverse was happening. The population continued to grow with no leveling in sight. Was Philip Ashmole's halo paradigm off the mark or were other forces at work on the Great Lakes?

The cormorants on Little Galloo Island, like other cormorant species around the world, are opportunistic feeders. As long as the fish that schooled close to the island were abundant enough and of sufficient size, they would fill cormorant bellies and allow the birds to feed close to home. More food, more fish brought back to the nests more often, meant healthier chicks. But what fish were these and where were they coming from? The fish that most noticeably answered that question was the alewife, one of the many species of herring, called sometimes the "sawbelly" in landlocked situations and known to biologists as *Alosa pseudoharengus*.

The convoluted saga of the alewife is a tale worth telling and in a real sense defined how the cormorant conflict evolved on the Great Lakes.

The alewife is one of about 180 exotic species that have invaded the Great Lakes by various means in the last two centuries. With apologies to all concerned, its common name, alewife, is supposedly derived from a comparison drawn between the fish's rounded belly profile and the rotund body type of a tavern owner's barmaid wife: an "ale wife." Regardless of its nomenclature, this herring, like its relatives, is an anadromous fish—a saltwater species that swims upstream and spawns in freshwater streams. After spawning, the adults, and later the resultant offspring, return to the sea to feed and grow until spawning occurs again the following spring. The difficulty arose when the Erie Canal opened in New York State in 1825, linking the Hudson River and Lake Erie. Hudson River alewives, seeking smaller streams in which to spawn, entered the Erie Canal system and found their way into the Great Lakes basin. There they adapted to a full-time freshwater environment and soon became permanent, reproducing residents of Lake Ontario. When the Welland Canal was constructed four years later, bypassing the natural obstruction of Niagara Falls, the ten- to twelve-inch silvery alewives spread easily to the other four lakes. By the 1950s, for an assortment of reasons, alewife populations jumped to incredible levels, with reports of alewives comprising up to 80 percent of total fish biomass in the Great Lakes.

Among the assorted reasons for the alewife explosion was the virtual elimination of the top predators from the Great Lakes, without which the alewife population grew unchecked. The lake trout, the popular hard-fighting sport and delicious food fish, was eagerly sought by anglers. This species is a long-lived trout with larger, legendary, hundred-pound specimens surviving twenty years or more, although Great Lakes anglers typically land much smaller fish. The lake trout's exceptional sport and dinner table popularity proved fatal to it, in that the species could not maintain its numbers against the prolonged campaigns of overfishing. Added to the annual fishing harvest was growing pollution, human development, and the invasion of the sea lamprey. Lake trout, having evolved in an ecosystem lacking anything even similar to parasitic sea lampreys, had none of the instinctive fear that would have helped them avoid lamprey attacks, and thus had no way to protect themselves from these eel-like parasites.

Adding more twists and turns to the saga, the alewife's own lifestyle and chemical physiology added significantly to the lake trout's downfall. Since alewives feed on zooplankton—the microscopic and near-microscopic animals drifting throughout the lakes, they also engulfed huge numbers of lake trout larvae, reducing the numbers in new generations before they had a fair chance to mature. And in a sort of bizarre cause and effect cycle, scientists found that the very act of lake trout eating alewives weakened the predator species. The herring's tissues contain a natural enzyme, thiaminase, a protein that breaks down thiamin, a vitamin B compound. The enzyme is absorbed by fish preying on alewives, and in lake trout the induced vitamin B deficiency of the adults was passed along to their eggs, which when they hatched produced weak, lethargic, disoriented larvae, which died off quickly.[6]

Commercial overfishing, parasitic fish, invading competitors, and metabolic disruption were too much for one species to bear. By the 1940s the lake trout was all but gone from the Great Lakes.

So the scene was set for the Little Galloo Island cormorant colony to find its place in the continuing alewife saga. The nonnative alewives invaded the Great Lakes system at roughly the same time as the lake trout population was under severe stress, and the major Great Lakes predator effectively disappeared. The alewife, because it faced no regular predation, became the dominant swimming life form in the lakes; its numbers soared for decades. Then the calendar rolled over to 1973, the year DDT was banned and cormorants were given another chance. As more and more cormorants survived, they encountered greater, not lesser, amounts of food surrounding their breeding grounds—practically unlimited, self-sustaining food. And conveniently, although alewives spend much of their lives in deeper parts of the lake, they move to shallower areas near shore when water temperatures reach the 55 to 60°F mark in the spring, precisely when Little Galloo Island cormorants need more food to feed their growing, hungry chicks. A study published in 1997 showed that in Lake Ontario in the early 1990s, and particularly on Little Galloo Island, the predominant food source for cormorants in the egg-laying and nesting stages was the alewife. As a percentage of fish species ingested, alewives composed upward of 90 percent of the cormorants' diet during the month of June.[7]

Another event that undid the application of Ashmole's model to the ecology of Little Galloo was the introduction of a second invasive species into the Great Lakes. The round goby, *Neogobius melanstomus*, is native to the Black and Caspian seas of Eastern Europe. It was first identified in the St. Clair River in 1990 and probably arrived through the pumping of ballast water from transoceanic freighters. The goby is a tough, aggressive, seven-inch invader capable of feeding on the bottom in total darkness, able to survive in polluted water, and, adding to its potency as a destructive exotic, spawns more often than the lakes' native fish. Today it can be found throughout all five of the Great Lakes. In the early 1990s round gobies were not even a measurable portion of the cormorant's diet. By 2003, alewives represented about 70 percent of the Little Galloo Island cormorants' intake.[8] By 2008 the round goby had filled the alewife's former niche, with over 90 percent.[9] Regardless of the particular forage species, it's clear that it was human activity, namely, canal building, the overfishing of predators, and the introduction of nonnative species through industrialization, that stoked the already vehement attitude fishermen held toward cormorants.

So, with new, massive, inexhaustible food supplies, the double-crested cormorant, having overcome near extinction through a twisted chain of events, defeated Ashmole's halo concept of limited population expansion and indeed moved into a period of unchecked growth.

The trail of Ashmole's Halo and the alewife saga was not a direct path, and yet the twists and convolutions continue. There is still the problem of billions of alewives and gobies, and what of the view of cormorants as symbols of ecological conflict? The ideas are linked, and not distantly.

In order to counteract the tremendous growth of the alewife and goby populations in the lakes, and to bolster the sportfishing industry, new and efficient aquatic predators needed to be injected into the unbalanced system. Two species of Pacific salmon, the coho and chinook, were first stocked in huge numbers in 1966–67. Earlier introductions were attempted in the 1870s, but the chinook did not reproduce in the lakes and died off. Hungry double-crested cormorants did take large numbers of them during the 1966 shore-stocking efforts, but with better protected stocking methods the hatchery salmon survived and grew. The chinook, which grows to forty-pound and larger trophies, and nicknamed "king salmon," became the angler's favorite target species on the lakes, with the smaller coho salmon running a strong second. The West Coast transplants, in fact, thrived and grew so well living on abundant alewife, and later goby, protein that within a few years new sport and commercial salmon fishing industries evolved, pumping fresh dollars into the economies of towns such as Henderson Harbor.

New revenues moved through the towns and harbors bordering the Great Lakes. The expanding commercial and sportfishing industry generated millions of dollars and supported thousands of jobs. It's at that point in time that cormorants were again looked on as competitive enemies, stealing income from fishermen and food from the mouths of their families. And why did this happen? It wasn't because Philip Ashmole's model was wrong. It was due to reasons that we will see repeated time after time, in other countries and on other continents: the elimination of top predators, the invasion of exotic species due to industrialization, and pollution and development of the waterways. And it's then that the shotguns were loaded in the countries of Europe and the village of Henderson Harbor in the United States, just a short boat trip from Little Galloo Island.

Young double-crested cormorants and adults on Little Galloo Island nests. *(Photo by Mitchell Franz.)*

Technicians preparing to spray cormorant nests on Little Galloo Island with corn oil to prevent eggs from developing. *(Photo by Mitchell Franz.)*

An "oiled" nest on Little Galloo Island.
(Photo by Mitchell Franz.)

Cormorants adapt their nesting behavior to whatever structures are available. *(Photo by Elizabeth Craig.)*

A close shot shows the interwoven sticks and twigs used to build cormorant nests. *(Photo by Elizabeth Craig.)*

A proud cormorant guarding its nest, looking less than sinister. *(Photo by Elizabeth Craig.)*

When threatened, a cormorant's strong parental drive motivates it to defend its nest and young. *(Photo by Elizabeth Craig.)*

Cormorants gather on the Hudson River waiting for tidal currents to drive forage fish through the nearby creek entrance. *(Photo by author.)*

A wet cormorant dries its wings after a pursuit dive for fish in a Cape Cod harbor. *(Photo by author.)*

A cormorant, usually tolerant of other bird species, protests the presence of this dozing gull. *(Photo by author.)*

With no attacks imminent, these alert cormorants take a lesson from Hitchcock's crows in *The Birds*. Younger, immature birds are pictured with lighter, grayish breast feathers. *(Photo by author.)*

On a Cape Cod marsh these cormorants relax on a swamped rowboat at high tide. *(Photo by author.)*

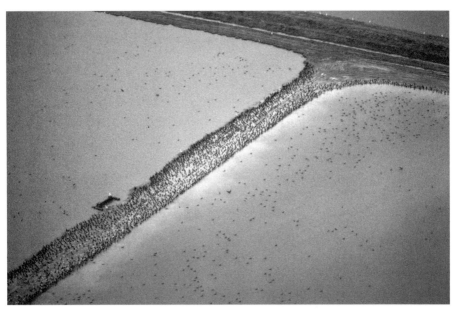

Thousands of hungry double-crested cormorants raid the waters of two Mississippi catfish ponds separated by a levy, which is also carpeted with cormorants. *(Photo courtesy Jerry Feist, USDA/APHIS/WS.)*

9

Little Galloo Island Revisited
Praise and Outrage

WHEN LAST WE LEFT Little Galloo Island, five men had turned the cormorant breeding colony into a shambles. New shotgun shell casings littered many parts of the ground. The newer casings could easily be distinguished from ones fired in earlier duck-hunting seasons by their gleaming, untarnished brass bases. Besides the shells, when the five men departed the island they left behind shattered nests and eggs and possibly close to a thousand dead, wounded, and dying cormorants.

The task of killing so many birds was not a neat and tidy operation. Unlike the duck hunters of seasons past, there was no shooting for sport or food. No hunter targeted a single bird in flight, led it, and fired. No hunter retrieved his quarry for the dinner table. That would be the sporting way. No. Men instead targeted nesting, roosting birds in the quiet of the night and shot at every cormorant they saw. And not just single birds but clusters of adults and flightless chicks. And not being sportsmen, these shooters had little or no respect for their quarry and left wounded birds to suffer their fate. And their fate was sealed by bodies peppered with shot, shattered wings, and missing or tattered legs and feet, torn away by the blasts. This "mission" was not ever to be compared to a successful sporting hunt. It was an ignoble, vengeful attack. But not everyone saw the shooting in that light but instead as a necessary act of desperation.

When word of the attack got around Henderson Harbor in the next few days, a feeling of relief spread through the town and through neighboring port and harbor towns as well. Charter captains and others had been complaining to the New York State Department of Environmental Conservation (NYSDEC) and federal agencies for years about marauding cormorants destroying local fish stocks. The captains felt their concerns fell on deaf ears and that nothing was being done to protect their interests. So it may have been the captains who visited the island that night. They took care of it themselves, regardless of the consequences. In semiwhispers the townspeople regarded the unknown perpetrators as local heroes. And as things typically happen in small towns, the identity of the shooters was probably a well-known, little-discussed secret, kept out of the daylight but surely having a quiet life of its own. But no secret is a secret forever, and the shooting was too large a secret, known by too many people in too small a town, to remain hidden for very long.

Part of the story can be told by following the postings of people's comments when the initial story became public knowledge.

> I've been pretty instrumental in trying to do this in a proper and legal manner. But everyone's been hearing rumblings about how people are going to go out and take care of the situation. Apparently someone finally wouldn't be talked out of it. (Ronald Ditch, fishing guide)[1]
>
> I only wish they'd killed every last one. (Ronald Ditch, fishing guide)[2]
>
> We've got no Kodak, no DuPont. You're either a fisherman or a farmer if you're going to live here. (Ronald Ditch, fishing guide)[3]
>
> The message has to be strong that people can't take the law into their own hands. (David Miller, executive director, New York chapter of the Audubon Society)[4]
>
> You can only oppress people for so long and they're going to strike out. (David McCrea, fishing guide)[5]
>
> Nature normally takes its course. But nature's out of balance right now. (Diane Gamble, marina operator)[6]
>
> This is a terrible crime. (Lt. Christopher Hanley, NYSDEC investigator)[7]

Everything here is a kind of weird balancing act between exotic species, native species, changes in the weather, changes in tourism patterns. The charter captains and the whole industry just has to adapt. (Stephen Fort, senior planner for Jefferson County)[8]

The Henderson Harbor area residents speaking out against cormorant incursions would not find themselves standing alone—in the summer of 1998 or in the past. From the time of the earliest European settlers cormorants had been perceived as a challenge to human fishing interests. The double-crested cormorant's tenacity and fishing prowess, coupled with its adaptability and sheer numbers, had established it as a worthy adversary and rival of sportfishermen in the first place. On Little Galloo the shooters never gave much, if any, thought to the cormorant's praiseworthiness. The five men did what they did in a brazen, self-proclaimed defense of their livelihood. There was also little doubt in the minds of reporters, prosecutors, and Henderson Harbor residents, those who either condemned or praised the island raid, that an understanding of the bird's evolution and perseverance would in any way have prevented that shooting. And as happens in war, when it's easier to kill the enemy across the battlefield from you when he's thought of as something less than human, it's easier for people to slaughter an animal when they see it as having no attributes justifying its existence—when in their minds they ask, "What good is it anyway?" If we view a species without considering what it does, how it lives, and who its relatives are, then we may have little problem with its extermination. But what happens after scientists and conservationists study the species and appreciate its uniqueness and some people still feel that the animal poses a threat to their interests? What do we do then?

The fact that cormorants are some of the top predators in their food chain never endeared them to humans either. Hunters and wildlife managers faced similar predicaments in the way they targeted other controversial predators, usually for one or both of two reasons. They either rationalized that the predators were competitors for what the hunters saw as prey reserved for them or the predators were envisaged as horrific, bloodthirsty, wanton killers of wildlife. Some predators with bounties on their heads—wolves, coyotes, grizzly bears, cougars, seals, great white sharks, and others—were ruthlessly hunted and persecuted to the edge of their reproductive limits. Unrestricted hunting policies accomplished their goals. These large predators began to disappear from their historic ranges. Continuous lethal culls of some species drove them to local extinction and in some cases nearly to national or global extinction.

In many cases these attempted exterminations backfired on the wildlife managers who had initiated them, throwing natural ecological relationships out of balance and producing more problems than they could ever anticipate. The indiscriminate killing of predators often reduced the genetic resiliency of their traditional prey; the prey grew weaker and more susceptible to disease as marginally strong individuals that normally would have succumbed to predation survived and produced genetically inferior progeny. Without the natural culling of prey by predators, populations, such as that of the whitetailed deer, grew to uncontrolled proportions, producing more deer-human conflicts (with gardeners, homeowners, and motorists) and a greater number of deer-to-deer interactions due to crowding in their habitat. The result was mounting winter starvation and increased transmission of parasites and diseases within deer populations.

Then, in a sort of falling domino or chain reaction other species were soon drawn into the unbalanced progression as well. Lyme disease, for example, a serious bacterial infection, is passed to humans from whitetailed deer through an intermediate host, the deer tick. This parasitic insect feeds on infected deer and transmits the Lyme disease pathogen to other deer and humans through its bite. Normal predation by wolves, coyotes, and other carnivores may have kept deer populations at natural levels, which in turn would have kept diseases from jumping from one species to another. The point being: let predators do what they do.

In the scheme of things, double-crested cormorants don't produce the same mental images as snarling wolves and coyotes, but they are, nevertheless, predators that interact with humans and fill a niche in the environment. Cormorant control and management policies, the culling of flocks and roosts, destruction of nests and eggs, may well be inevitable, but they may also have their own unintended, unpredictable consequences. Government shooting programs, the itemizing of the bird's perceived nuisance aspects, and reports of its fish stock pilferage not only damage the cormorant's "reputation" but threaten a delicate and necessary balance in the environment, a concept the shooters on

Little Galloo Island eventually came to misunderstand, along with the nature of the cormorant itself.

Of all the expressions and phrases that its human opponents used to depict the double-crested cormorant, including the most derogatory ones, the term *quitter* was never one of them. Despite massive attacks on their numbers, driving a graph of cormorant population levels into deep valleys, they persevered. Native Americans on both the Pacific and Atlantic coasts harvested plentiful cormorants from their huge isolated colonies, devouring their eggs with relish. Recent archaeological discoveries of shell mounds on the West Coast reveal that prehistoric natives, perhaps as early as twelve thousand years ago, exploited the abundance of waterbirds, both onshore and offshore. Before these findings, archaeologists pictured these ancient Indians as living in a perfectly balanced environment where they reaped what they needed and nature replaced what they consumed, and at the same rate. This view proved to be a bit of a fantasy. As the populations of Indian settlements grew, they consumed much of their surrounding wild crops and local game. They then moved their hunts to isolated island rookeries, harvesting more and more waterbirds to supplement gaps in their diet. One target species was the cormorant, particularly its eggs and chicks. Researchers have postulated that the Indians' relentless bird hunts depleted these populations to the point of causing local extinctions.

Very early European visitors to the Pacific coast found the same low, harvest-depleted numbers of waterbirds. Historical records show that these explorers mingled and traded with local Indians and in the process introduced European diseases such as smallpox, for which the Indians had no natural defense whatsoever. Disease quickly decimated the tribes. But with the Indians' unfortunate destruction came relief to the cormorant island rookeries, whose populations soon recovered when the seasonal hunting excursions ceased. Later permanent European settlers arrived on the coast where they met the huge repopulated flocks of fish-eating double-cresteds, which they saw as a challenge to their colonial success.

Unlike the native peoples, European settlers were not substantial consumers of cormorants or their eggs, which may have had something to do with how they named the bird. Double-crested cormorants were originally called sea crows, a name derived from the Latin terms *corvus,* meaning "crow," and *marinus,* meaning "marine." Combined through

French-language usage the term *cormorant* evolved. So, despite a need to exploit their natural resources for food, the settlers were much less likely to dine on what they considered the local version of the familiar crow.

These settlers from the Old World brought their fishing traditions with them and grew to depend on fishing local waters for much of their sustenance, and perhaps for the life of the entire settlement. Even though cormorants were not thought of as menu items, the settlers did see the great colonies of these fish-eating birds as their enemy: competitors for the same fish. Consequently, they slaughtered huge numbers of cormorants in what they considered a defense of their fish stocks and, ironically, sometimes used cormorant meat to bait their fishhooks. As a persistent species, cormorants returned year after year, though in reduced numbers, to fish the same waters alongside their new human rivals, even as the persecution continued.

Returning to Henderson and the summer of 1998, reports of the shooting on Little Galloo Island rang out from many quarters. The national media, sensing the newsworthiness of what many people outside the Henderson area thought of as a barbarous act, kept the story alive, at least in print, for more than a year. The story sold papers. To urban newspaper readers, including armchair naturalists, serious conservationists, city birdwatchers, Central Park pigeon feeders, sympathizers, and animal lovers of all types, it was difficult to imagine what motive would drive someone to wantonly kill so many wild birds. Today, more than ten years after the incident, there exists an amazing "paper trail" on the Internet. Following accounts in the *New York Times* archives, one can trace the thread much like the visual devices film directors use to quickly move their stories along. Visualize a scene in which a massive newspaper press runs at blazing speed, spitting out headlined editions, fading to the same press at a later time spinning out more papers, flashing updated headlines to advance the story. Or picture a police detective tacking a series of newspaper articles to his wall or bulletin board to watch the progress of his case. In looking at our case, the July 9 shooting on Little Galloo, we can pin the following articles and dates on our own bulletin board.

"A Slaughter of Cormorants in Angler Country" (August 1, 1998)
"The Dead Cormorants of Little Galloo Island" (August 8, 1998)

"In a Fishing Hamlet, No Grief for Cormorants" (August 9, 1998)
"Grand Jury to Investigate Killing of Cormorants" (September 1, 1998)
"9 Men Plead Guilty to Slaughtering Cormorants to Protect Sport Fishing" (April 9, 1999)
"Fishermen Fined and Sentenced in Killings of 2,000 Cormorants" (August 12, 1999)

The titles tell the story—almost. The *Times* articles lay out the sequence of events, but it's the details, quotes, and emotions that really tell the story of what happened on Little Galloo and afterward.

The first article begins the progression. A day or two after the deaths of the cormorants, biologists and assistants working on Little Galloo Island studying the birds' diet discovered the dead and dying adults and chicks. The state biologists "encountered heaps of carcasses of fledgling cormorants, piles of shotgun shells and starving chicks squawking weakly amid the carnage."[9] This had to be a very disheartening sight for anyone who cared anything about wildlife. Before any suspects were even considered for the crime, officials intuitively felt that it was surely an outgrowth of the long-standing disputes between wildlife conservation and commerce on the Great Lakes.

In 1998, the Henderson Harbor area supported sixty professional captains working out of nine marinas. Charter boat skippers, guides, and their paying clients were kept busy pursuing Pacific salmon and other sport fish; about a quarter of the chartered and guided trips targeted near-shore populations of smallmouth bass since these fish prefer rocky habitats, as opposed to deep, open-water areas. A New York State Bureau of Fisheries report detailing a Lake Ontario fishing boat census recorded 84,671 fishing boat trips in 2004, which couldn't have been too far off the 1998 figure.[10] The fishing industry was still active, and money was being made.

Near shore usually means shallow water, and on Lake Ontario that also means cormorants. The diet studies discussed earlier showed some cormorant consumption of game fish in parts of the season, but other fish species composed larger portions of their meals throughout the season. But again, with the smallmouth population in decline and cormorant numbers on the upswing, the birds were pegged as the culprits.

In Henderson Harbor, conservation versus commerce conflicts

more times than not centered on reducing cormorant numbers, but since the early 1970s double-cresteds had been protected from hunting by the Migratory Bird Treaty Act. In 1998, Little Galloo Island was the breeding-season home to as many as 8,400 breeding pairs of cormorants and their active nests.[11] Federal wildlife biologists estimated that cormorants in the eastern basin of Lake Ontario, which includes Little Galloo Island and Henderson Harbor, ate between 400,000 and three times that number, 1.2 million, smallmouth bass every year.[12] Without a doubt, that's a lot of fish. But these fish clearly were not the catchable-size smallmouths sought by anglers; rather they were mostly fingerlings and juveniles. Therefore, in terms of young smallmouth bass, the biomass drawn from the ecosystem by cormorants was only a small percentage despite the magnitude of the figures. The birds did, however, reduce the *absolute* number of bass by perhaps 1.2 million a year, which again was a small percentage (less than 1.0 percent) in relation to the total population of billions of smallmouths. Cormorants clearly were not the only factor in the declining smallmouth fishery. So once again perception outpaced reality.

Verlyn Klinkenborg, a *New York Times* editorial observer, wrote the second article in the headline sequence. In his essay, Klinkenborg developed an insightful analysis of what happened on Little Galloo and why. It's worth our time and effort to grasp the concepts presented in his short six-hundred-word piece. The subhead, "Contemplating the Cost of Ignorance and the Ethic of Sport Fishing," reveals the point of the article. The *Times* author, a professed fisherman himself, supposed that the shooting on Little Galloo Island, like most environmental or ecological crimes, was an economic crime. He made the case that "humans could fish Lake Ontario with devastating efficiency . . . so efficiently, in fact, that it would soon doom cormorants and all other fish-eating birds almost as an afterthought." Klinkenborg's concept of "devastating efficiency" can be accepted easily as either a predictor or a historical review. It not only could happen; it did happen. Recall the few short decades it took the Great Lakes' commercial fishing industry to *devastate* Lake Ontario's native Atlantic salmon to the point of extinction and the industry's virtual elimination of sturgeon and lake trout.

Klinkenborg pointed out that sportfishing, to be rightfully called a sport, must be "intentionally inefficient." Hooking and landing a fish is not a given. The fact that a sporting angler fishes bait on a single hook, uses a breakable line so the fish must be "played," or casts a single lure gives the fish the choice to bite or not to bite, and once hooked, it allows the fish a fighting, sporting chance to escape. The author also commented that because of the added-value services provided by captains and guides, the sport fish itself had become a powerful economic force, defined and measured by the dollars fishermen spent to catch it. And so we can conclude on our own that with economic forces come economic crimes.

One last and telling concept Klinkenborg discussed in the relationship among fish, sportfishermen, and cormorants, brought to light by the island shooting, is the direct competition between humans and other species, a struggle not often acknowledged since so few people today live in nature and recognize our historic and biological ties to it. The article concludes with a succinct statement: "When humans compete with other species, other species always lose."

Other species always lose.

When we look again at the residents of Henderson Harbor and surrounding areas, we should remember past citizens who, in the War of 1812, refused to allow military or naval operations to be launched from their village. The men, and women, did put up harsh resistance to the British and their Indian allies when their lakeside village and interests were threatened. The town had an unambiguous history of opposing the authority of a distant government, British or American. Andrew C. Revkin, author of the third article in our *Times* list, explained that "local residents resent what they regard as the intrusion of unreasonable regulations written far away and long ago."

The harbor's present residents may not be directly related to its historic inhabitants, but their independent spirit, American "rugged individualism," and an affinity for dissent certainly come through. Revkin maintained that in and around Henderson Harbor the shooting was looked on much like the Boston Tea Party. Others, like John P. Cahill, then New York commissioner of environmental conservation, saw the Little Galloo Island raid not at all as a protest but as "an act of savage brutality."

A protest is a way to voice a complaint or frame an objection. It's a demonstration of opposition. Punctuating dissent with shotguns and slaughtered wildlife is something else. The brutality of the "protest" exposed itself when a wildlife assistant, Laura Brown, stepping among the bright, spent, twelve-gauge shell casings "found 7 freshly killed birds. Then 7 led to 200, and 200 more."[13] The sight of dozens on dozens of rotting, protected wild bird carcasses had to have registered as a far more vicious scene than a few soggy cases of British tea drifting across Boston Harbor. The two deeds were simply not equatable.

The fourth article in the *Times* list pointed out, as the investigation progressed into discovering the identities of the shooters, that it seemed that many Henderson Harbor residents knew but would not tell authorities who the individuals were. Even a reward of $50,000, offered by the owner of the Houston Rockets basketball team, Leslie Alexander, did not elicit any useful information. Chris Henley, an investigator for the NYSDEC, said, "Henderson Harbor is a small town and people who live and are rooted there may know, but $50,000 may not be enough to relocate, which they may feel they would have to do if they talk."[14]

Reward or no reward, relocation or not, investigators from state and federal agencies spoke to many people around Henderson Harbor. The historic and continuing conflicts between charter fishermen and the US Department of Agriculture, USFWS, WS (Wildlife Services), NYS-DEC, and other wildlife management bureaus naturally boosted the charter captains to the top of the investigators' suspect list. The anti-cormorant comments from the captains published and republished in newspapers like the *Times* had to help focus the investigators' eyes on the outspoken skippers.

While ballistic experts studied the shells for fingerprints and brands of shotguns used on Little Galloo Island, a federal grand jury was scheduled to meet in Syracuse, New York, in the first week of September 1999 to look into the killing of the protected birds.

The case had put a lot of newspaper ink in circulation, inflaming public sentiments, so both investigators and prosecutors were determined to follow the story through to a conclusion. In January 1999, Mitch Franz, a Henderson charter captain, was subpoenaed to appear before the grand jury for the second time. He testified that he knew nothing about the shooting. After the grand jury session he was approached repeatedly and questioned by investigators. They even tracked Franz down to where he was legally hunting from a duck blind to ask more questions. By that time he knew the investigators by name and had no ill feeling toward them in that "they were doing their job, and I was doing mine."[15]

In a long and convoluted sequence, Franz appeared a number of times, received a grant of immunity signed by then US attorney general Janet Reno, and still refused to name names. Another captain, however, under intimidating pressure from prosecutors, agreed to testify before the grand jury. All of the ten men identified by that captain, except Mitch Franz, who was never questioned again, were called to the grand jury to testify.

Apparently, what legal pressure and rewards could not accomplish the power and intimidation of a federal court building and grand jury subpoenas could. The truth came out. Grand jury proceedings are secret, conducted behind closed doors and illegal for prosecutors to reveal, but the Syracuse panel evidently produced results, since, as related by the *Times'* fifth and sixth articles, five men pleaded guilty to killing the double-crested cormorants on Little Galloo. David McCrea, Ronald Ditch, Mitchell Franz, Louis Cook, and John Kabot did the actual shooting that night; Rex Allen and Angelo Aversa later hid the guns; and Aversa's and Ditch's three sons admitted to shooting cormorants in other, smaller incidents.[16] If any of the names sound familiar, it's because they were cited earlier in the sequence of quotes.

Federal law at that time provided sentences of up to a $5,000 fine and six months' imprisonment for each count (each protected bird) in violation of the Migratory Bird Treaty Act. Maximum penalties could have amounted to millions of dollars in fines and years in federal prison, but, probably as a result of a negotiated plea as opposed to a jury trial, the five shooters were required only to serve six months of home confinement, pay a fine of $2,500, and contribute $5,000 to the National Fish and Wildlife Foundation. Mitch Franz lost his immunity because he refused to testify before the grand jury, and, threatened with a federal felony charge of conspiracy, a conviction of which meant the loss of his captain's license issued by the Coast Guard, he agreed to be subjected to the same penalties as the other shooters on Little Galloo.[17]

The others involved in the Little Galloo incident and earlier, related shootings received similar but lighter sentences. The combined crimes resulted in the deaths of nearly two thousand federally protected birds. And with the sentencing on August 12, 1999, and a statement from the assistant US attorney who helped prosecute the case, Craig A. Benedict, that "Vigilantism, and especially this kind of cruel killing by individuals, has no place in resolving a complex population issue,"[18] the case was closed.

The government's Little Galloo Island case was indeed closed, but double-crested cormorant issues and conflicts on the Great Lakes were far from ended. It behooves us to recall the historic mind-set of Great Lakes fishermen. Commercial fishermen overfished and depleted one species after another, and yet they fought regulations claiming that quotas, limits, and other restrictions would surely ruin their industry. Instead they ruined it themselves by fishing to the last catchable fish. The attitude of sportfishermen is somewhat closer to that of the conservationists in that they realize that in order to preserve their sport they have to avoid the "tragedy of the commons" and leave some fish for the next guy and the next generation. But what some Great Lakes anglers are not willing to concede is that they have to share the resource with other species, namely, the double-crested cormorant, which uses fish as its food source.

A prime example of this philosophy flows from a group called the Great Lakes Sport Fishing Council (GLSFC). In a message from the president posted on the GLSFC website shortly after the Little Galloo Island incident, Dan Thomas, a longtime proponent of stringent cormorant control and culling, makes the claim that "there was nothing inhumane about killing those birds" and cormorants "seem to do more harm to the web of life than any perceived good."[19] His boisterous ideas hark back to earlier, less enlightened times when men offhandedly decided which species deserved to survive and which did not and whether one species was simply better than another. The really frightening claim Thomas makes is that "those fish stocks being eaten by these fish eating marauders belong to our fishing partners, our fishing friends, our children and their children's children. They are damaging our property—our precious and not unlimited resources."[20] Our property? Who do the fish stocks belong to? Under Thomas's logic any species that needs to share a natural resource with humans is to be considered a pilferer and an enemy and therefore a species undeserving of life.

Another tactic the GLSFC uses to denigrate cormorants is the toxic effect of the runoff from their droppings. Apparently, high levels of mercury, PCBs and DDT, and their derivatives had been found in waters surrounding cormorant colonies on the Great Lakes, again including Little Galloo Island. In a December 1999 posting on their website, the GLSFC blamed "gluttonous" double-crested cormorant populations for these chemicals leaching into local waters during seasonal rainy periods.[21] In the minds of the members of the group the cormorants therefore posed a threat because they consumed fish contaminated with man-made chemicals unknown in nature, as well as ingesting dangerous metal toxins released into the lakes as industrial effluents. Again, it is far easier to name the cormorant, a middleman of sorts in the toxin procession, as the culprit than it is to get at the real root of the problem: man-made toxins in the lakes.

Holding the birds accountable for the pollution also takes the emphasis off the fact that many of the fish caught and brought home for the dinner table by paying clients of Great Lakes charter captains are contaminated with the same toxins. The 2009–10 *Official Fishing Regulations Guide,* issued with the purchase of a New York State fishing license, contains several health advisories for fish taken from Lake Ontario. The advisory for lake trout over twenty-five inches and brown trout over twenty inches is "eat none." For chinook salmon, rainbow trout, and smaller lake and brown trout the advisory is "1 meal/month." Also "women of childbearing age, infants and children under the age of 15 should not eat any fish from the waters listed above [Lake Ontario]." Thus, citing cormorants as the source of the pollution, the council, in a defensive move perhaps, made the contamination seem less ominous and the fish more attractive to their paying customers. Their tactics again attempted to portray the cormorant as an innately sinister species best subjected to frequent control and culling programs. Since cormorants are so successful at what they do, let's blame the birds for the pollution, too.

Cormorants, like many other black birds, are not among the chirpy little birds people love so much; they were despised and hunted. The cormorant has been hated throughout history. Cormorants were said to be unfit for eating in the Old Testament and were compared to Satan sitting on the Tree of Life in Milton's "Paradise Lost." A fourteenth-century painting shows archers in Venice trying to shoot the birds in the back.[22]

And in modern times, New York State and Lake Ontario were not the only sites on the Great Lakes whose interests posed conflicts with cormorants and whose residents would prefer to see them dead. In places such as Green Bay, Wisconsin, on Lake Michigan, residents voiced mixed opinions about double-crested cormorants ranging from "nothing but a flying rat" to the "majesty" of large flocks of them foraging.[23] During the first few years of the current century the USFWS proposed and later authorized what it refers to a Public Resource Depredation Order. Original proposals ranged from doing nothing at all to establishing hunting seasons or open seasons on cormorants. Even though control measures were not necessarily excluded by some conservation groups, the idea of a hunting season for cormorants wasn't very appealing either since, unlike waterfowl such as ducks, there is no use for the resource. Dr. Lee Martin, director of the St. Lawrence Bird Observatory in 2000, relates why hunting seasons for cormorants are not practical: the birds are inedible; because they fly far from shore, they are not readily accessible to hunters; and, unlike other waterfowl, they are not easily drawn in by floating decoys.[24] Hence, we see no holiday recipes for cormorant à l'orange.

These depredation programs, permitted under the Migratory Bird Treaty Act, allow not only federal agencies but states and tribal authorities to use lethal force, under permit, to limit and control the growth of bird populations that are said to damage private and public recreational and commercial interests. In simpler terms, on the Great Lakes, it allows federal, state, and tribal wildlife officers, with specific permits, to kill substantial numbers (thousands or even tens of thousands) of cormorants when it's felt they eat too many fish. The numbers of birds taken must be reported to appropriate federal agencies, and individual citizens are not sanctioned to shoot the birds. Leaving the reporting of "birds taken" to those who do the taking created a natural loophole in the regulations. In a 2002 newspaper article in the *Detroit Free Press*, author Eric Sharp relates a passed-along anecdote about how a fish hatchery biologist was issued a federal permit to kill fifty of the many gulls raiding his hatchery. In his office, the biologist stored about a thousand shells for the facility's twelve-gauge shotgun. When asked if he had so much ammunition because his workers were such bad shots, he answered, "Oh, we're good shots. We just don't count very well."[25] True or not, the story highlights a popular sentiment about one way to solve one problem.

The Little Galloo Island incident and the justifications for the shooting of so many birds, including the destruction of trees and vegetation, parallel much of what was happening across the border in Canada. The dispute in Canada over cormorant culling was and is a well-followed issue, not unlike many of the cormorant disputes in the United States. In 2004, the *Toronto Star* published an article concerning the shooting of cormorants over the issue of dead trees with the clever subhead "Tree Story Just Doesn't Fly."[26] The article claims that the birds' crime was *not* that their guano destroyed the trees in which they nest but that they had "made enemies in the lucrative sport fishing industry"; thus, the double-crested cormorant's plight was a tale of money versus nature and "humankind's inability to tolerate any other species that it finds remotely bothersome." The author, Thomas Walkom, went on to say that cormorants, as big black birds that hang around in gangs and defecate a lot, have few friends in the human world. But aside from killing trees, like many other colonial, tree-roosting birds do, their real sin is that they eat fish. Our mantra from earlier in this chapter, "When humans compete with other species, other species always lose," is recalled by Walkom in his own way: "All creatures play a key role in the balance of nature. You can't pick and choose . . . which ones are shot." The Ontario provincial government authorized the killing of nearly seven thousand double-crested cormorants to protect the trees of Presqu'ile Provincial Park so that other, less competitive, fish-eating birds might destroy them instead.

Another issue raised by anticormorant interests is the topic of disease. Double-crested cormorants are carriers and victims of Newcastle disease, which can also infect poultry, other wild birds, and humans. It was first identified in cormorants from the St. Lawrence River in 1975. The disease is transmitted either through bird guano or by humans who have been in contact with infected birds, primarily poultry and laboratory workers.

Newcastle disease is caused by specific strains of RNA viruses attacking the digestive tracts and respiratory and nervous systems of certain domesticated birds, including chickens and turkeys. The infection is highly contagious and capable of inflicting nearly 100 percent mortality rates in affected flocks of birds. Outbreaks in 1990 and 1992 in the United States and Canada caused the deaths of tens of thousands of birds, mostly cormorants. Human infection is uncommon and usually associated with direct worker contact with infected birds or their droppings. People infected with the Newcastle virus can contract conjunctivitis, the inflammation of the covering of the eyeball, mild fever, and general flulike symptoms, which may last three to five days. Proponents of cormorant shooting sometimes use Newcastle disease as a footnote to their rationalization for culling programs, but science is not on their side: only one report of Newcastle disease transmission from cormorants to poultry has ever been documented, and no extensive mortality has been reported in other wild birds sharing local environments with infected cormorants. So again cormorants were charged through emotional perception but in reality found not guilty.

When it comes to dealing with avian pests the topic of shooting always turns up as one of the alternatives. The terms *lethal force, lethal measures,* and *culling programs* tend to leave readers and listeners alike with a picture in their minds of a violent but controlled and necessary action that is over when the shooting ceases. The difficulty with maintaining that compassionate image is that, like many circumstances in nature, a culling action is not as "clean" or simple as we would like to picture it. When shooting birds in flight, afloat, or nesting, not every hit is a kill, not every fallen bird is dead, and not every wounded bird dies quickly. A Canadian conservationist group, Cormorant Defenders International (CDI), whose mission is to end cormorant persecution, posted a pair of disturbing videos on their website depicting the aftermaths of two government-sanctioned culls in provincial parks. Hundreds upon hundreds of cormorants were killed to maintain what was said to be the natural growth and integrity of the parklands. The videos show birds wounded by government officers and left to die. Much like the illegal vigilante action on Little Galloo Island, hunters ignored injured birds with legs blown off and badly maimed in other ways and simply included them in their approximated kill counts. The culls were not clean, not humane, but again, one way to "solve" one problem.

Solving one problem in one way doesn't necessarily mean that a particular situation is resolved forever or that it hasn't created new conflicts to face immediately or in the future.

The destruction of trees and other vegetation by cormorants on their breeding islands certainly raised the hackles of foresters and park managers, probably not without some cause. They saw the trees as permanent, fixed features on the islands and sought to preserve and defend them in their natural habitat. The fault in their accepted wisdom is that the habitat itself was changing. In the Vermont/New York border waterway Lake Champlain, on a six-acre Vermont-owned island, the birds destroyed "a lush preserve of basswood, elm, green ash, and cottonwood trees" that were once host to a variety of other fish-eating birds and songbirds.[27] Canadian newspapers reflected the same problems in their article titles: "30,000 Cormorants Destroying Lakeside Park" and "Shooting Cormorants over Dead Trees Raises Suspicions about Liberal Motives . . ."[28] The stories rang with descriptive phrases, including "wintry apocalypse,"[29] comparing the cormorant's destructive power to the ravages of a nuclear attack.

But the question needs to be considered of what interests are actually protected by shooting wild birds to preserve foliage. No doubt exists that double-crested cormorants create what can be considered visual eyesores in areas where they congregate. Breeding cormorants break off leaves and branches to construct their large, flat nests. Tons of what scientists refer to as "nutrient transfer," acidic guano deposited by the birds, chemically burns leaves and leaches into the soil to eventually kill the entire tree in a period as short as three years. But as in the "eye of the beholder" adage, the barren landscape vision to the human eye is only an uncomfortable sight to those expecting a cozy, pastoral, manicured, parklike setting. It may not be our mind's perfect image, but it is the result of natural processes. And for our adaptable cormorants the remodeled islands suit their purposes just fine.

In an idyllic world this view of nature might be justified, but nature doesn't work within the same parameters as a park naturalist. If outlaw loggers moved onto the islands and cleared the woodlands illegally they might have a more substantial line of reasoning. But that's not the case here. Nature changes, climates fluctuate, animal and plant populations shift, grow, and fade. Aggressive species in both kingdoms overwhelm less assertive species. The natural processes that we observe are not guided by any anthropomorphic sense of fairness akin to a human conscience. Some groups of organisms are simply more successful in a certain environment than others.

That is not to say that humans have no right to step into a situation in an effort to level the biological playing field for species under stress. But we have to realize that at times the field is tilted to favor human interests. Or at other times efforts can be misdirected, misguided, or heavy-handed, producing unintentional negative emotions in the public at large and perhaps working against the very species the agency is trying to protect. Take, for example, what takes place every year on the beaches of Cape Cod in Massachusetts. The piping plover, a cute, fragile shorebird, arrives each spring on Cape Cod beaches to nest directly in the warming sand. Perfect camouflage plumage renders the plover chicks virtually undetectable to the human eye, even at close range. But the same cannot be said for the senses of predators prowling the beaches or the skies above them. The eyes of hawks, crows, and gulls key in on the shadows and movements of vulnerable chicks in and around exposed nests. On the ground, skunks, raccoons, and coyotes comb the shore, testing and tasting the air with their senses, looking to make easy meals of plover chicks and eggs. And unlike other birds, such as the double-crested cormorant, plovers on the Cape do not have a strong instinct to renest, to lay a replacement brood of eggs, when the original nest and brood are destroyed by storms or roving predators. The adults shift out of their reproductive mode into their postreproductive lifestyle, skipping a reproductive season. As a protective measure authorities controlling the national seashore close long stretches of beach to fishermen, sunbathers, and even kite fliers. The closures put smiles on the faces of conservationists, but beachgoers turn their anger toward the fragile birds with a popular bumper sticker reading "Piping plover: Tastes like chicken."

The predator-prey relationship is part of the natural ecology of the seashore, and species of gulls, aside from their garbage-dump raids, also threaten nesting plovers. At around the same time as the Little Galloo Island incident, the USFWS decided to reduce the gull population on Cape Cod beaches near Chatham to protect the nesting plovers. The plan was to kill about six thousand gulls by baiting them with bread spread with butter laced with a poison specific to the gulls. The idea was that the gulls would die quietly on the beach, out of sight, out of mind. The gulls, however, did not follow the USFWS game plan. They did indeed eat the poisoned bread, but not all of the targeted gulls consumed the full deadly dosage in one sitting and did not drop dead on an isolated stretch of beach. Instead, many gulls ate only a portion of the bread and, feeling poorly, flew inland in search of freshwater rather than saltwater. A number of gulls, while flying over the quaint village of Chatham, finally succumbed to the poison. Gulls dropped from the sky onto roofs, parked cars, and the streets filled with tourists and their children. Local newspapers carried accounts of dying, flopping birds being euthanized by humane veterinarians. USFWS agents had to go into hiding due to death threats. All in all, the operation did not go well.

These and other protective measures, when they work as planned, do guard the vulnerable nesting birds, but what they do not do is allow for the plover's ability to meet its own challenges. By killing the unevenly tempered gulls and stopping children from flying kites on the beach, policymakers remove the natural impulse the plovers have to move to a more isolated, safer beach to nest or at least learn to distinguish a child's kite from a marauding hawk or gull. Without developing their own adaptations to changes in their environment—increased predation, increased human activity—animals such as piping plovers are doomed to living in the little scenarios created by their human protectors as a substitute for defining their own place in the environment.

Getting back to the conundrum of the dead trees, a reporter for the Canadian Broadcast Corporation, Sharon Oosthoek, in a 2009 essay explored the ideology of protecting one part of the ecosystem over another. It centered on Parks Canada's five-year culling plan to protect a Lake Erie island's rare Carolinian forest from the unloved double-crested cormorant. *Carolinian forest* is a term used by Canadian foresters to describe zones of deciduous, broadleaf trees that include a mix of oak, hickory, chestnut, ash, birch, and walnut, which provide shelter and food for any number of wildlife species. Middle Island lost more than 40 percent of its forest canopy to cormorants, and the remainder of the timber faces a similar fate. Oosthoek wrote, "While officials hope culling the habitat-altering cormorants will save an ecosystem that makes up just one per cent of the country, others grapple with the ethics of re-engineering nature." She adds that it is a conflict between "the arrogance of thinking we can reassemble ecosystems and those who say that in some cases, we have no choice."[30] Oosthoek also cites Parks Canada's chief ecosystem scientist, Stephen Woodley, in his explanation of why this forest in particular is worth preserving. It seems that the Carolinian forests in Canada represent strains of the tree species existing at the limit of their geographical range. These trees have a distinct genetic constitution not seen in organisms living in the center or their range. These "edge populations" have adapted to the more rigorous northern conditions, and their unique genes represent a reservoir of features ensuring continued biodiversity. What Woodley fails to discuss is the vast adaptability of the double-crested cormorant and the genetic strength it took to overcome the destructive effects of DDT and its toxic cousins. It again appears that the cormorant's own success is at the heart of man's attempt to "reengineer" its place in the ecosystem.

A journal article by Linda Wires and Francesca Cuthbert from the Department of Fisheries, Wildlife and Conservation at the University of Minnesota described the major fallacy in the reasoning of agency administrators and wildlife managers attempting to regulate the powerful drives of the double-crested cormorant.[31] It addressed the unrelenting joust between one faction, the fishermen, aquaculture interests, and agencies, and the second, the conservation groups, squeezing the cormorants in the center.

> Although managing cormorants may benefit some fisheries and resolve human-cormorant conflicts, setting population objectives for cormorants based entirely on fishery or other objectives derived from human values significantly departs from the concept of conserving birds in natural numbers and natural habitats. Given the global state of ailing fisheries and increasing government sensitivity to wildlife acceptance capacity, self-regulating populations of cormorants may only be possible in cormorant "safe zones" where human interests are not allowed to influence cormorant numbers. Because few such "safe zones" are likely to exist, those committed to the conservation of fish-eating birds should vigorously oppose strategies that set population objectives based entirely on human interests rather than species biology and regional ecology. Finally, we urge the avian conservation community to support broad conservation strategies based on ecosystem health and process that recognize humans, fish and cormorants as three components of a complex system driven by many species and dynamic interactions.

Change is inevitable. Human industry, taken in its broadest, most creative meaning, changes the world, as does nature itself. Attempts to ignore these changes or battling changes on the wrong fronts produce programs dedicated to preserving an "absolute image" of what nature should be, what it should look like, and how it should meet continuing human expectations. This absolute vision produces sanctioned culling paradigms, clandestine vigilante bird massacres, and human confusion over which species are cute and desirable and which species are not as cute, less desirable, and thus more easily killed.

The shooting on Little Galloo Island was probably the most publicized attack on cormorants, but people other than Mitch Franz and the Henderson charter captains apparently also took offense at cormorants eating fish to survive. About 80 percent of Big and Little Charity Islands, located in Michigan on Lake Huron's Saginaw Bay, are owned by the USFWS. Two years after the incident on Little Galloo Island, on May 30, 2000, agency biologists found a second slaughter of breeding double-crested cormorants. Investigators determined that more than five hundred cormorants had been illegally killed on Little Charity Island. They found dead birds scattered all over the five and a half acres of the island; some were shot while sitting on their nests. The story of Little Charity Island is short because there is little to tell. In 2000, the year of the shooting, the investigation was said to be continuing, but no one has ever faced charges for the crime.

Similar to these two episodes, it was the earlier, massive shootings, which occurred before such actions were even crimes, that prompted the enactment of international wildlife treaties, the passing of related legislation, and the establishment of regulating and controlling agencies, the topics discussed in the next two chapters.

PART 3

Cormorants & the Law

10

Treaties and Legislation
War of the Wilderness

[I]t shall be unlawful at any time, by any means or in any manner, to pursue, hunt, take, capture, or kill, possess, offer for sale, sell, offer to barter, barter, offer to purchase, purchase, deliver for shipment, ship, export, import, cause to be shipped, exported, or imported, deliver for transportation, transport or cause to be transported, carry or cause to be carried, or receive for shipment, transportation, carriage, or export, any migratory bird, any part, nest, or eggs of any such bird, or any product, whether or not manufactured, which is composed in whole or part, of any such bird or any part, nest, or egg thereof.[1]

THIS LEGAL PASSAGE was written and designed to specifically protect wild migratory birds from unrestricted hunting, collection, and persecution throughout their international range. Before 1918 no such federal or international protection existed. The quote is from US Code Title 16, chapter 7, subchapter II, known less formally as the Migratory Bird Treaty Act (MBTA) of 1918. The enactment of the MBTA in its time would create a powerful tool for protecting migratory birds, but until then conservationists, policymakers, and legislators still had much ground to cover.

Doubtless, the saddest thing about the passage quoted here is that it had to exist at all. Today most people, though not aware of the MBTA itself or its unambiguous legal clauses, would consider unthinkable the shooting of majestic wading birds, the sale of their skins and feathers, and the listing of colorful songbirds and their eggs on the menus of popular upscale restaurants. Like the protection afforded by the treaty, this modern sensitivity did not always exist

In the 1880s it was quite possible to see a fashion-conscious socialite walking Boston's streets or New York's avenues wearing a hat decorated with the stuffed skins and feathers of an entire family of songbirds, which had been shotgunned out of their nest. An online *Audubon* magazine article further described the nineteenth-century millinery fad: "America's hat craze was in full swing. In the 1880s trendy bonnets were piled high with feathers, birds, fruit, flowers, furs, even mice and small reptiles. Birds were by far the most popular accessory: Women sported egret plumes, owl heads, sparrow wings, and whole hummingbirds; a single hat could feature all that, plus four or five warblers."[2] This wealthy social maven may not have been so much coldhearted and bloodthirsty as she was unknowing, and perhaps uncaring. The men who created, and the others who copied, European Elizabethan fashions, thought little of the wildlife they destroyed to sustain the feather industry. But in fairness to the women who wore their products, the merchants did formulate a pattern of lies and misrepresented their trade practices to their clientele. In an effort to conceal the cruelty of their business they falsely let it be known that the feathers and plumes were painlessly collected from the nests of molting birds. Unsuspecting shoppers were more than willing to accept these unreliable explanations because they wanted to, and because they had little or no firsthand knowledge of the true "collection" process.

The hunting of birds just about anywhere, for any reason, was virtually unregulated. The collection of feathers for the millinery trade "took a tremendous toll—200 million wild birds per year by some estimates."[3] Some of the most sought after feathers were the long, delicate plumes, the aigrettes, of egrets and herons, produced by the birds during their breeding seasons to attract mates. The adult birds were killed on their nests and harvested, leaving chicks to starve and eggs to be scavenged by predators. *Tons* of plumes, with each feather weighing a mere quarter ounce, representing the deaths of hundreds of thousands of wild birds, reportedly passed through the salesroom of just one London feather merchant in a single year.[4]

Stories abound in the literature of the widespread killing of birds. Market hunters on the Chesapeake Bay killed up to fifteen thousand canvasback ducks daily. Bobolinks, cedar waxwings, and wild birds' eggs were served in restaurants. One writer described a woman's dress adorned with the skins of three thousand Brazilian hummingbirds; Bald eagles' eggs were offered to egg collectors in catalogs for ten dollars and house wrens' eggs for fifteen cents.[5] Feathers from terns and gulls, as well as from various raptors, sparrows, Baltimore orioles, warblers, tanagers, owls, bobwhites, flickers, and double-crested cormorants, turned up in many markets and on an assortment of women's fashions, particularly their stylish hats. Even today's Christmas Bird Count, the annual methodical count of all birds and listing of all avian species found within designated "count circles," is the conservationists' outgrowth of the traditional Christmas Day "match hunts." The match was built around hunting teams that shotgunned *everything* in sight, birds, mammals, whatever, and then competed to determine which team's pile of carcasses at the end of the day was larger. The result of all this wasteful killing was that bird populations across the country and around the globe simply crashed.

One of the early voices to be heard protesting the pervasive killing was that of George Bird Grinnell. Grinnell was far more than a nature enthusiast; he was a true and authentic outdoorsman. A complete inventory of his accomplishments might well be mistaken for the lifelong deeds of several men. Grinnell was an American writer, historian, anthropologist, and naturalist. He authored or coauthored twenty-six books, many concerning the history and culture of Native American tribes, including the Cheyenne, Blackfoot, and Pawnee. He developed such a close relationship with the Pawnee that they adopted him into the tribe with a given name, White Wolf. As a naturalist, he traveled with Lt. Col. George Armstrong Custer during his 1874, thousand-man-plus, exploration of the Black Hills of South Dakota, an expedition taken up prior to Custer's fatal mission to the Little Bighorn. Grinnell's intense interest in the American West moved him to lobby Congress in the successful preservation of the nearly extinct American buffalo. Back east, he helped organize the New York Zoological Society, now known as the World Conservation Society. As editor and contributing natural history writer for *Forest and Stream* magazine, Grinnell told the story of the slaughter of millions of birds for the feather trade. In 1886 his efforts expanded to founding in New York of the first Audubon Society.

Grinnell's writings incensed and invigorated the public, but with his energy stretched only so far and the society's finances stretched beyond recovery, the movement collapsed—and the shootings continued.

Picking up where Grinnell left off, a number of Boston's wealthiest and most influential women joined together under the leadership of Harriet Hemenway, whose family was known for its social and ardent abolitionist activism. In a succession of afternoon teas, Hemenway and her cousin, Minna Hall, recruited other women in their social circle in an attempt to boycott feather- and bird-skin-adorned hats. Among their ranks were listed the names of Elizabeth Cabot Agassiz, first president of Radcliff College, and Sarah Orne Jewett, a popular novelist of the time. Together these women eventually organized and founded the Massachusetts Audubon Society. As a cultural necessity of the time, they drafted a male, William Brewster, in 1896, founder of the American Ornithologist's Union, as the masthead president of their fledgling conservation group.

The Audubon women, practicing what we would call "networking" today, managed to coordinate the efforts of many disparate women's groups common in the turn-of-the-century women's club movement, the majority of which had little to do with conservation, let alone birds. From that network they built a sizable grassroots force that exerted pressure on policymakers across Massachusetts. They spoke at schools, worked with teachers, distributed literature, and sold bird identification charts and calendars to spread the concept of preserving wild bird populations and boycotting products of the brutal feather trade. At the same time, their better-known male club officers lobbied legislatures and pressed their case with business leaders, as well as urging colleagues within scientific circles to become active in the Audubon movement.

One of the first federal outgrowths of the Audubon Society's education and lobbying programs was the Lacey Act, the oldest national wildlife protection legislation in the United States. The Lacey Act, more than a century old, is still one of the most powerful tools available to law enforcement officers fighting illegal wildlife trafficking. At its outset, the act was directed at protecting game and wild birds from poachers and feather-market hunters, but has been amended and expanded several times to include amphibians, reptiles, and certain invertebrate species. In 1988, a provision was added to the act regulating the activities of guides and outfitters, who did not conduct illegal hunting themselves but organized and conducted illegal hunts for their paying, shooting clients. As recently as 2008 the Lacey Act was amended to include coverage of plants, exotic woods, and goods manufactured from them such as furniture and even guitars.

The act is named for the Iowa congressman, John Lacey, who proposed the legislation in the US House of Representatives in the early part of 1900. Lacey, representing large numbers of midwestern farmers, sought to protect agricultural interests through the preservation of important native bird species that made their living eating millions of crop-damaging insects. Lacey's bill also included provisions enabling the Department of Agriculture to *reintroduce* songbirds that had been eliminated from locales by excessive hunting. The legislation was also designed to prevent the introduction of destructive exotic species and to augment existing state laws frequently ruled by courts as "insufficient" to counteract the illegal interstate and international sale and transportation of wildlife and their products. The act made it illegal to poach wildlife in one state and sell it in another. An environmental crime that would normally be considered a poorly enforced misdemeanor in each of two states could be prosecuted as a federal felony because the plundered wildlife products had been transported across state lines. The final bill also contained powerful clauses prohibiting the falsification of documents related to the shipment of wildlife products, carrying both civil and criminal penalties. In addition to the Department of Agriculture, the Department of the Interior and Department of Commerce would be responsible for enforcement through their individual agencies, such as the USFWS. President William McKinley signed the Lacey Act into law on May 25, 1900.

After the enactment of the Lacey Act and the election of Theodore Roosevelt to the presidency in 1900, "the bully conservation champion,"[6] the concept of the preservation of wildlife species took root in America. By 1901, thirty-six states boasted local Audubon associations, which united to form the National Association of Audubon Societies in 1905 with William Dutcher, George Bird Grinnell's ally in the first abortive Audubon efforts, selected as its first president.

Using funds donated and dedicated for the specific purpose,

Dutcher initiated a program supporting a network of game wardens patrolling wading bird rookeries along the length of the Atlantic coast, including Florida's massive Everglades. Some of the enlisted wardens, working for thirty-five dollars a month, such as Guy Bradley, were former feather-market hunters themselves, aware of the locations and practices used by current illegal hunters. In 1908, Bradley was killed in the process of arresting a poacher. And unfortunately Bradley was only the first; two other wardens were killed in the same year in similar situations, a sign of the money to be made, even at the lowest echelons of the feather trade.

The Lacey Act did a great deal to fortify state statutes but presented problems in federal enforcement and prosecution due to a lack of funds and personnel, with as few as six game wardens employed for federal conservation enforcement. In 1913, Congress responded to hundreds of thousands of letters and telegrams protesting the continuing commercial feather hunts by passing a package of bills protecting birds and virtually ending the US feather trade. In 1913 the Weeks-McLean Act, named after its sponsors in the US House and Senate, prohibited the hunting of migratory birds on their spring breeding grounds and the resulting marketing of their feathers. Placing its authority in the Department of Agriculture, the secretary created the first national closed seasons for the hunting of migratory birds. The act's effectiveness, however, was short-lived. The new laws were soon challenged on a constitutional basis with the claim that they violated the Tenth Amendment, which states that "the powers not delegated to the United States by the Constitution, not prohibited by it to the States, are reserved for the States respectively, or to the people." The avian protection laws were set aside by several courts, which ruled that Congress had no constitutional right to pass laws on matters directly related to state issues. The feather trade was again free to continue its destructive market hunts.

The formulation of federal legislation to protect birds faced a major setback but was not defeated. The US Constitution does indeed limit the power of the federal government to make laws restricting activities in and of the states, *but* it also empowers the federal government to enter into treaties with other nations to protect the interests of the United States and its citizens. The way around the prohibitions established by the courts was to internationalize controls and regulations aimed at curbing the feather trade, in short, a treaty. In the United States, constitutional law allows the negotiation of treaties, which can be passed by a two-thirds advice and consent action by the Senate. Final ratification and approval takes place when the president affixes his signature to the final document.

To protect our migratory birds, the United States entered into negotiations with Great Britain, which then represented Canada, that country having not yet gained its full independence from England. The Migratory Bird Treaty with Canada, also known as the Convention between the United States and Great Britain (for Canada) for the Protection of Migratory Birds, was signed by President Woodrow Wilson and ratified by both nations in 1916. The treaty constructed a uniform system of protection and preservation for certain species of birds that migrated between the United States and Canada and were deemed harmless or beneficial to man. It also created seasons closed to hunting and prohibited the taking of insectivorous birds, protecting valuable agricultural crops. It also laid the groundwork for the sanctioned killing of birds proven to be threatening to agricultural interests. The skeleton of the international treaty, its wording and provisions, stood neatly on its own, but putting muscle on the bones to enable the day-to-day enforcement of the treaty within the United States required implementing legislation, namely, the Migratory Bird Treaty Act of 1918, quoted at the opening of this chapter.

Under the MBTA it became illegal to "pursue, hunt, take, capture, or kill" egrets, owls, eagles, swallows, gulls, herons, and terns. The original treaty with Canada provided international protection for hundreds of bird species representing many taxonomic families.

One species not safeguarded by the treaty was *Phalacrocorax auritus,* the now familiar double-crested cormorant. In the early 1900s cormorant population numbers in the Great Lakes were not at all what they are today. Colonizing, nesting cormorants from the US Midwest moved eastward, and the birds adapted well to conditions on the Great Lakes; their numbers swelled through the 1940s until DDT and its cousins poisoned the food chain, nearly destroying the cormorant's annual reproductive potential. At the time of the MBTA enactment, the second decade of the twentieth century, few cormorants nested in the Northeast; therefore Canadian wildlife officials were reluctant to classify cormorants as migratory birds and offer them the treaty's protection. But by the early 1970s, when DDT was finally banned, cormorant populations had already been destroyed by the persistent toxic contaminants. Without protection, the few remaining double-cresteds could easily be permanently eliminated by a few seasons of overhunting and persecution. But that, too, could be managed through convention negotiations and amendments.

The initial 1916 treaty with Canada was the first of four bilateral treaties between the United States and other nations with similar concerns for preserving their bird populations. Mexico was the next nation to sign on, in 1936, Japan in 1972, and the former Soviet Union (Russia) in 1976. The MBTA was amended after each convention to adjust US laws, incorporating the new bilateral agreement. The treaties and the act were periodically updated several times to keep the accords current. It was in one such 1972 revision to the agreement with Mexico that double-crested cormorants were first included in provisions of the MBTA, at last adding them to the list of protected migratory birds.

The contaminant-threatened, nearly extinct double-crested cormorant was now protected by federal statute throughout the United States and in the southernmost sectors of its migration. In Canada, as stipulated by the original bilateral treaty, cormorants were still not protected under federal law. Added to the ban on persistent insecticides, the new prohibition on the indiscriminate killing of cormorants gave *P. auritus* at least an earned fighting chance of recovery.

Other successive international conventions and treaties directly or indirectly came to the aid of cormorants and other migratory birds and added to their chances of longtime survival. One of the greatest threats to these birds was, and continues to be, the destruction of valuable habitat. For wading birds and pursuit divers this habitat means marshes, wetlands, and what came to be the term that caused real estate developers to sneer when they uttered it—*swamps.* Verbally and conceptually it is easier to justify plowing under or filling in a "worthless swamp" than it is to raze a productive "biodiverse wetland." One of the two accords that began life in the 1960s is the Convention on Wetlands of International Importance, known as the Ramsar Convention after the city in Iran where initial negotiations took place.

The Ramsar Convention is unique in that it is the only worldwide environmental agreement that deals with a specific type of ecosystem: wetlands. Ramsar affords an international platform for the protection and wise use of wetlands and their wildlife. It was negotiated through the decade of the 1960s, signed in Iran in 1971, and finally enacted in 1975. The convention employed the broadest characterization of the word *wetland*. Few types of aquatic and marine habitats were excluded in its provisions. International protection was thus granted to specific "lakes and rivers, swamps and marshes, wet grasslands and peatlands, oases, estuaries, deltas and tidal flats, near-shore marine areas, mangroves and coral reefs, and human-made sites such as fish ponds, rice paddies, reservoirs, and salt pans."[7] The "wise use" goal of this convention tied human treatment of wetlands to the concept of sustainable development. Wetlands could and would be protected. Though indirect, the Ramsar Convention, through its international coverage of all manner of wetland habitats and their resources, created one more layer of protection for the unloved, migrating double-crested cormorants, including cormorant species other than *P. auritus* found around the globe.

The second international conservation agreement negotiated through the 1960s and ratified in the 1970s was the Convention on International Trade in Endangered Species of Wild Fauna and Flora (CITES). The convention was originally endorsed by 80 countries in 1973, this time in Washington, DC; present-day membership includes about 175 nations. At the core of its mission statement is the concept of protecting endangered species and those approaching that status from the exhaustive international trade in approximately thirty thousand plant and animal species, either as live specimens or "products derived from them, including food products, exotic leather goods, wooden musical instruments, timber, tourist curios and medicines."[8] These resources may also appear in markets as fur coats and dried herbs. Many of these organisms were exploited at levels that posed severe threats to their very existence, particularly when coupled with factors such as habitat destruction and environmental pollution.

Cormorants did not appear on the species lists of this accord, but the fact that, in the 1960s and early 1970s, eighty governments sat down to expressly limit and prohibit the expanding international trade in market-valuable wildlife and plants gave hope and brought attention to birds like cormorants, which had little commercial value but were so blatantly persecuted by humans and so heavily poisoned by widespread toxins that they had little "wiggle room" left in which to survive.

11

Agencies and Wildlife Conservation at Work

Cormorants in a Vise

THE DOUBLE-CRESTED CORMORANT numbers about two million across its entire range, where they find trouble at all four points of the compass and several in between. The conflicts, aside from relatively minor ones centered on vegetation and competition with other birds, invariably focus on fish. And fish and fisheries are always about economics—money. Commercial fishing interests see direct connections between natural predation on fish by wildlife, cormorants for one, and their income; sportfishermen see cormorants as affecting the enjoyment of their sport. In situations where money competes with wildlife, the result consistently involves control and management of the wildlife. How do you control and manage two million resourceful cormorants? The US government has a number of ways it goes about deciding what it considers appropriate control measures. The various conservation groups, also included in the mix of stakeholders, look at waterbird conservation from several other angles, including the feeling that agencies like the US Department of Agriculture (USDA) have a strong bias favoring agricultural interests over wildlife protection. Where does that leave our cormorants? Centered directly between these two powerful forces. That's the story that follows.

International treaties are federal documents enforced on the federal level by federal officials. As agreements between sovereign states or international organizations, they depend on international law for their underlying structure and enforcement. The guiding law for treaties is the Vienna Convention on the Law of Treaties, ratified by over a hundred countries and also recognized by nonsignatories as binding as a restatement and expression of customary international law. The Vienna Convention essentially serves as the "Windows Operating System" of international legalities, the program that runs under the others and keeps them all in line.

On an international scale, the Vienna Convention acts as a treaty guideline, but in the United States these treaties live much of their lives in individual states with federal enforcement providing a common factor. So, depending on the issues involved in a specific treaty, implementation is assigned to a particular Cabinet department of the government, typically the Department of the Interior or the USDA, in questions of wildlife or crops and plants, respectively.

Various agencies within the two departments then set about designing protocols for enforcement across the country. Within the Department of the Interior, the US Fish and Wildlife Service (USFWS) is responsible for wildlife regulation enforcement. The second department, the USDA, oversees APHIS, the Animal and Plant Health Inspection Service, an operating unit whose stated mission is "to protect the health and value of American agriculture and natural resources."[1] A unit division of APHIS, Wildlife Services (WS), "is responsible for providing federal leadership in managing problems caused by wildlife and provides assistance to agencies, organizations, and individuals in resolving wildlife damage problems on public, tribal, and private lands."[2] It's clear, even to the legal layman, that wildlife and agricultural issues, which typically tangle into chaotic knots on their own, are further complicated by the overlapping, conflicted, and intermingled responsibilities, protocols, and policies of federal agencies and departments.

Wildlife Services has an interesting and mixed history of its own, and understanding its current actions is at least partially dependent on a look at its background as an agency. Under a series of different titles and umbrella agencies, usually including the phrase Predatory Animal and Rodent Control (PARC), WS was moved in 1939 from the Division of Game Management to the Department of the Interior. In 1980 the agency became Animal Damage Control and under pressure from agricultural interests was moved back to USDA in 1985, where it remains today as Wildlife Services.

The agency itself underwent a number of convoluted changes in its mandate and mission. Its early goals under the Animal Damage Control Act, signed into law in 1931 by President Herbert Hoover, involved bringing under control a wide range of predators and rodents from mountain lions, wolves, and coyotes to prairie dogs and gophers. Its resulting, long-standing programs of poisoning predators with such potent toxicants as strychnine and cyanide brought it much criticism and public outcry, particularly in the "Leopold Report" issued in 1963. The committee that published the report, headed by A. Starker Leopold, son of the noted conservationist Aldo Leopold (discussed later), reported that the agency was excessive in its application of poisons and had caused the indiscriminate deaths of wildlife other than the predators and rodents. The report did lead to a series of lesser changes, including the establishment of an external advisory panel and, of course, another name change. The agency's history is also plagued by any number of audits and investigations having to do with missing quantities of poisons, storage compliance issues, the training of its personnel, and unsecured agency aircraft and fatal aerial accidents, all of which, after the attacks on September 11, 2001, became substantially larger issues for the agency.

Overall, WS's current stated vision, "to improve the coexistence of people and wildlife,"[3] may reflect a new attitude toward wildlife, but nevertheless still carries the baggage of mismanagement, a card often played by conservation groups.

Every US treaty is formulated to eliminate potential or actual international disputes. By its very nature, and regardless of how well a treaty is crafted, its enactment inevitably limits, regulates, or promotes some American financial interest over others; otherwise there would be little point to the treaty in the first place. And because treaties are often imperfect and lacking in one detail or another, treaty negotiators invariably include policy exceptions to allow agencies to smooth the imperfections. The twisted policy dealing produced an interesting conundrum for US wildlife policymakers. What this management poser meant to double-crested cormorants was that they were protected from hunting and persecution by revisions of the MBTA except when they were legally *not* protected from hunting and persecution by the MBTA.

The Department of the Interior's wildlife enforcer, the Fish and Wildlife Service (USFWS or FWS), at its inception in 1871, was created by Congress, during the first term of President Ulysses S. Grant, as the US Commission on Fish and Fisheries. After the Civil War it was initially charged with finding new food fish for rapidly expanding cities and chose the Asian carp, a decision, as we'll see later, that backfired. Today the FWS has as its mission in broad terms "working with others to conserve, protect and enhance fish, wildlife, plants, and their habitats for the continuing benefit of the American people."[4] When cormorant populations rebounded after the EPA banned DDT in 1972, and the birds' status shifted from being a nearly extinct species to a nuisance, the control of this new pest fell to the FWS. It was then that the agency faced conflicting undertakings. It was now mandated by its mission and international treaty to "protect and enhance" some species and maintain biodiversity, while at the same time was held responsible for reducing the numbers of other species through regular lethal measures, including culls and other campaigns.

The working tool of the FWS for controlling cormorants and other nuisance species was the PRDO, the Public Resources Depredation Order. The MBTA, as amended in 1972, added the cormorant, along with almost every other bird not covered in earlier versions, to its protected list. The PRDO, on the other hand, allowed the FWS to permit lethal control measures as exceptions to the MBTA's protections. In order to apply lethal controls to cormorant populations a procedure had to be in place to systematize the action.

The USDA's Wildlife Services program was the first government contact made in the depredation request process. Landowners or public agencies could apply to WS, expressing their particular wildlife problem, which in our case centers on double-crested cormorants. Officials at WS then evaluated the detailed application and signed off on it. The FWS then confirmed that the application included the numbers of cormorants to be killed, an estimate of the economic loss that would be sustained if the birds remained, an assurance that this culling would not endanger the species as a whole, and a statement that the action carried no objections from state or tribal authorities. If lethal controls were included as options in the plan, then a FWS migratory bird permit office would issue a depredation order authorizing the killing of the problem-causing migratory birds. Even then, the applicant was expected to continue to incorporate nonlethal techniques along with the approved lethal methods program. And since shooting strategies rarely provided total solutions to wildlife conflicts, the applicant was also obliged to describe alternative long-term actions that could be used to eliminate the cormorant problem, even though the lethal permit had already been issued.

The late 1990s was a time when it was thought that cormorant numbers were at or had exceeded the all-time historic high of more than a million birds. In 1998, the FWS announced a ruling that would allow catfish farmers and other aquaculture facility managers in thirteen states to kill cormorants preying on their ponds. The decision was not intended as a method of controlling the *overall* cormorant population but as a site-specific relief for fish farmers combating cormorants feeding on commercially valuable fish. The birds could be taken only at facilities certified by APHIS, and aquaculture managers were required to retain logs detailing the numbers of cormorants killed. Again, FWS officials needed to walk that very narrow line between defending treaty-protected cormorants and permitting legal hunts to cull fish-eating birds.

As a partial explanation of their "preserve versus control" conundrum, the FWS concluded some of its lethal permit text with a disclaimer stating that it was "both a leader and trusted partner in fish and wildlife conservation, known for our scientific excellence, stewardship of lands and natural resources, dedicated professionals and commitment to public service." As the "primary statutory authority" for managing US migratory bird populations under the MBTA, and taking its commitment seriously, the FWS tops some of its web pages with the motto "We make it happen." From a philosophical viewpoint these statements leave readers with a softened, but not unfounded, sense of the power that the authoritative agency wields.

In reading the procedures and guidelines just discussed, it would not be unrealistic for the average citizen to fall into a hypnotic state characterized by slowly closing, glazed-over eyes. The "alphabet soup" of acronyms alone is capable of inducing a deep slumber. And what has been mentioned thus far is an ultrasimplistic compilation of the various government texts. To accomplish anything, APHIS, WS, and the FWS had to work with each other and with fifty state governments, their agencies, and the many public and lobbying organizations within their borders. As a means of organizing and prioritizing how these federal agencies address environmental issues Congress passed the National Environmental Policy Act (NEPA) in 1969.

The act established the powerful and underlying idea of balance and harmony between humans and the natural resources and ecosystems of their environment. It "requires federal agencies to integrate environmental values into their decision making processes by considering the environmental impacts of their proposed actions and reasonable alternatives to those actions."[5] The legislation requires the use of interdisciplinary approaches by federal agencies in their planning. The NEPA process is overseen by the Council of Environmental Quality (CEQ) and its staff, which monitor environmental quality and evaluate federal environmental programs.

The process detailed by NEPA became the template that guides the actions of all federal agencies when their actions might affect some element of the environment. Any action covered by NEPA must undergo an Environmental Assessment (EA) to determine if the undertaking will have a significant effect on the environment. If the action will have little or no environmental influence, then a Finding of No Significant Impact (FONSI) is issued, allowing the plan to proceed. If, however, the EA indicates a significant environmental impact, the agency must prepare an Environmental Impact Statement (EIS), which mandates a far more detailed analysis of any environmental effects, including input and comments from the public and other stakeholders. All parties are then permitted to comment on the resulting Draft Environmental Impact Statement (DEIS) before the Final Environmental Impact Statement (FEIS) is completed and entered into the *Federal Register.*

Even with computers, maintaining and inputting records, conducting inspections and wildlife inventories, and handling follow-ups can be seen as clogging an already overwhelmed system. And as cormorants expanded their breeding grounds, recolonized historic sites, and colonized new ones, conflicts with human activities increased in very much the same proportion as the growth of their population. In order to manage the barrage of conflicts and complaints concerning cormorants, the USFWS, with WS, developed a "selective management alternative" in 2003, which contained two depredation orders that specifically addressed the continuing cormorant damage issues. The two orders deal with the separate realms of cormorant raids: *public* and *private* resources.

The Public Resources Depredation Order was established to reduce actual cormorant numbers and/or lessen the risk of damage to public resources, such as free-swimming fish stocks and those fish raised in federal, state, and tribal hatcheries and slated for later release into public waters. The order also sought to address the cormorant's competitive effects on other colonial nesting birds, guano destruction of plants, and the reported devastation of habitats. At the heart of this highly expanded order was the fact that it empowered officials to keep the numbers of cormorants affecting human interests in check without the issuance of federal permits.

The scope of the PRDO encompassed twenty-four states located predominately in the South and Midwest, with Vermont and New York thrown in as boundary states.[6] The order was a boon to fish farmers and sportfishermen in that it would defend wild and cultured fish stocks against actual or perceived cormorant threats to catfish and sport fish species. The depredation action not only alleviated a great deal of federal paperwork, but the ability to assist local industries transferred from FWS federal authorities to state agencies came with another offering: the financial responsibility to do so.

State budgets often involve a great deal of infighting and battles among legislators to reach a final agreement. States like New York are known for extended deliberations, many times overlooking statutory deadlines for the passage of state budgets. And when times are lean state representatives often look more closely at how their constituents' taxes are spent, meaning cormorant-control dollars might suddenly disappear from a budget. But in some states, such as Michigan and Wisconsin, the cormorant's apparent interference with fish and fishing, thus affecting tourism and the view of these states as vacation destinations, was seen as a severe drain on the economy. As recently as 2007 Michigan state representative Steven Lindberg, in his support of a package of state bills passed to control cormorants, wrote, "The cormorant is one of the biggest threats to our environment and our special way of life in the Upper Peninsula. Many people come here to fish our lakes and rivers. If cormorants continue to plague our waterways and destroy our fish populations, it will have a devastating effect on our economy."[7] To back up the crusade against the cormorant "plague," the bills also created a "cormorant control fund" to finance the programs paid for with money or assets from a variety of sources.

In a state like Wisconsin, as one of the twenty-four states referenced in the PRDO, and one well known for its fish and fishing opportunities, it would not be unexpected to find cormorant-control funding items in its budget. And just how much does it cost to manage and shoot cormorants?

One particular problem area in Wisconsin is a number of small islands in Green Bay, at the mouth of the Fox River, which were colonized by cormorants after DDT was banned in 1972. It was commonly thought that fish-eating cormorants had reduced the numbers of many fish species sought by recreational fishermen. Wisconsin's 2005 budget included an estimate of about fifty thousand dollars for each of the five Green Bay islands for the first four years of the control program. The state would also contract with APHIS/WS to supply sharpshooters to shoot enough cormorants to hold the island populations at what it considered manageable levels. So Wisconsin allocated two million dollars (plus the cost of contracted sharpshooters?) to kill an unspecified number of wild migratory birds, which might or might not have been *actually* responsible for the declining fish stocks in Green Bay. How much research, and how many surveys and studies, would the same two million dollars finance to determine the extent and effects of overfishing, pollution, and development? No one would know because those are three very unpopular alternatives, which many legislators and regulators might not be ready to explore. Sometimes, it's just easier to shoot cormorants.

The second depredation order, the *private* version, known as the Aquaculture Depredation Order (AQDO), was designed to allow aquaculture producers and state and federal hatchery operators to shoot cormorants in order to protect fish stocks from actual or *potential* cormorant raids on their property, again without federal permits. This time, thirteen states were included, some of which were also authorized in the public version.[8] All of these states were located in the South, with the exception of Minnesota, a midwestern state with aquaculture interests. Shootings were limited primarily to pond and tank sites themselves and did not include roost sites off the fish farmer's property.

Through the creation of these two depredation orders the FWS reduced or removed much of the "red tape" required to openly shoot double-crested cormorants, with little or no actual accountability aside from the keeping of unsupported shooting logs. The USFWS wielded its enormous regulatory power in an attempt to unify the varied rules and regulations governing how states and organizations dealt with double-crested cormorants. Agencies, as arms of the government, are born to create and enforce policies, but their actions are not always welcomed by those affected by said policies. We saw earlier how the USFWS faced the conflicting mandates of protecting cormorants and culling them at the same time. And it is important to remember that Wildlife Services acts as an arm of the Department of Agriculture: protecting the interests of agriculture, farming, and farmers. Its mandate is not necessarily to protect wildlife but rather to be "responsible for providing federal leadership in managing problems caused by wildlife."[9] "Civilian" conservation organizations outside the immediate circle of government agencies also recognized this federal conflict of goals and sought to capitalize on it, challenging the FWS's scientific pronouncements, its legal obligations, and particularly the lethal regulations derived from them.

The initial depredation order issued in 1998 produced an estimated "take" of 46,664 cormorants per year for the 1998–2001 four-year period. The revisions to the order would enable the taking of far more cormorants with far less accountability.

The killing of more than 185,000 protected wild birds in that four-year period angered conservation groups, but they were further incensed by proposed amendments to the depredation order expanding the justifications and scope of the authorized shootings, predominantly at and around aquaculture sites. The amendments now added the allowable "taking" of cormorants by WS personnel at roost sites in the vicinity of aquaculture sites and the control of cormorants to protect fish and wildlife on public and private lands in specified states. Cormorants at roost sites were to be taken only by shooting with firearms, while birds preying on public resources (free-ranging fish) "may be taken by egg oiling, egg and nest destruction, cervical dislocation [neck wringing], firearms, and asphyxiation."[10] All in all, it was a stunning menu of choices in the continuing pressure put on double-crested cormorants to stop being cormorants.

One of the many conservation groups stepping up to argue against the action, the American Ornithologists' Union (AOU), established a panel of five scientists to review the USFWS's 2001 Draft Environmental Impact Statement on Double-Crested Cormorant Management. The five authors, affiliated with a number of universities, produced a thorough, detailed report on the draft, as well as the final EIS filed by the USFWS.[11] In a realistic attempt to evaluate cormorant versus human conflicts, the report discussed perceived problems, the biological reality of them, and the USFWS's plans to reduce ongoing conflicts. The review of the proposal reported that the USFWS's final and draft EISs advocated the killing of 160,000 and 250,000 birds per year to resolve existing cormorant conflicts. The USFWS provided no means of assessment for success or explanation of the estimates.

Through the citation of more than sixty sources, most of them primary research papers, the panel made the point that much of the scientific evidence presented by the USFWS in its EISs was flawed or strongly skewed in favor of financial-based stakeholders, namely, the commercial fishing and sportfishing industries. The six reasons the authors furnished certainly raised a large number of questions concerning the validity of the agency's actions. According to their report:

1. The scientific evidence supporting the proposed action was weak;
2. The analysis of the data was simplistic;
3. The management plan proposed by USFWS was inadequate and had a poorly evaluated potential to be effective;
4. The consequences of the proposed action on cormorants were unknown and appeared to be punitive instead of mitigatory;
5. The assessment of success was unclear; in the DEIS, success was based on public perception and not on scientific results. The FEIS was not clear on how success would be assessed; and
6. There was no adequate mechanism for monitoring the population effects of the plan, nor for deciding when to terminate management actions.

In the process of developing its EISs, the USFWS proposed what it saw as the six viable alternative management options for cormorants. The options were labeled A through F, and over time each "letter" became a well-known symbol standing on its own as a choice for stakeholders and respondents to the proposal. The agency described the six choices in the *Federal Register* of October 8, 2003.

A. Described as no action, no change, or status quo meant maintaining the current policies of limited numbers of double-crested cormorant (DCCO) depredation permits, including both lethal and nonlethal measures.

B. This nonlethal management alternative, directed mostly at hatcheries and aquaculture sites, "would not allow the take of DCCOs or their eggs." Harassment and techniques of physical exclusion would be the only measures permitted.

C. Under this alternative local damage control would be increased, "but agencies and individuals would still have to obtain a depredation permit."

D. The PRDO alternative, the one chosen by the USFWS, provided for increased killing *without requiring permits* by authorizing "state fish and wildlife agencies, Federally recognized Tribes, and APHIS/WS to take DCCOs found committing or about to commit, and to prevent, depredations on the public resources of fish (including hatchery stock at Federal, State, and Tribal facilities), wildlife, plants, and their habitats." Alternative D also applied to commercial freshwater facilities, meaning catfish farms in the South and baitfish suppliers in the Midwest.

E. This option "would reduce regional DCCO populations to pre-determined levels." Committees in undetermined regions would develop individual population goals for cormorants based on the "best available data," which would be reached through lethal and nonlethal controls "carried out at nesting, roosting, wintering, and all other sites in order to achieve those objectives as rapidly as possible."

F. The regulated hunting alternative would create open seasons and bag limits for hunters wishing to shoot cormorants. This choice was discarded because it was thought that few hunters would be drawn to the idea of cormorants as quarry, probably because the birds have no value as table fare or trophies.[12]

The same *Federal Register* documentation discussed the responses to the proposed ruling received during the public comment period. The USFWS received about 9,700 letters, e-mails, and faxes, of which 85 percent opposed the proposal. The agency appeared to dismiss the overwhelming opposition expressed in the letters, "the vast majority of which were driven by mass email/letter campaigns promoted by nongovernmental organizations." The USFWS, by law, sought and reported negative responses to the proposal. They were reported but apparently largely ignored. Agency officials evidently thought little of them since they were generated through citizen-supported conservation groups and others outside their own circles.

In the *Federal Register*, the USFWS shared more than fifty comments and its response to them. Several of the comments focused on the USFWS's obligation to protect cormorants, not kill them. The agency's responses typically described methods for the killing of cormorants as "humane, but lethal techniques," its measures as "approved by the American Veterinary Medical Association," and shooting as "a legitimate and effective technique." The term *effective* is not necessarily synonymous with *humane*. Far from it, as one may ultimately exclude the other.

Another group of comments centered on the idea that the Fish and Wildlife Service had ignored biological evidence and caved to special interests in formulating its updated depredation proposal. The agency responded with phrases and statements such as "an appropriate balance of the public interest" and "[S]ocial, political, and economic factors contribute to the Service's decisions." In these answers the USFWS blatantly ignored the 85 percent opposition responses while it mentioned balancing the public interest. The decision also conveniently overlooked the many scientific reports and diet studies that showed fish-eating cormorants having very little effect on overall valuable fish stocks. The AOU review also stated, with citations, that "studies that report depletions of desirable fish by cormorants only quantify fish numbers at local spatial scales (i.e., only a subset of a biological population), or are based on small samples."[13] The review also pointed out other instances where small amounts of data were extrapolated and generalized to strengthen claims of widespread fishery damage caused by cormorants. When blamed for scapegoating cormorants for fishery decline the agency referred to "some cases" in which cormorants were a significant contributing factor to the decline and added, without explanation, that cormorant management "is likely to have beneficial impacts."[14] The five authors of the review discussed the many other factors relevant to declines in fishery production, adding that referring to cormorants "as the cause of these perceived problems" was not justified by the science reviewed in the USFWS's FEIS.

Also quoted in the AOU committee review was the statement in the FEIS itself that the commercial fishing industry had "experienced a steady decline for reasons unrelated to fish-eating birds."[15] When challenged by such comments, the USFWS again recognized other causes of the decline of fisheries but stated that a "comprehensive analysis of these myriad factors is outside the scope of the EIS."

Included in the "myriad factors" were no doubt the natural ecological processes of habitat alteration, competition, and environmental shifts. A comprehensive report released by the National Marine Fisheries Services (NMFS) of the National Oceanic and Atmospheric Administration (NOAA) discussed the importance of ecological considerations.[16] It opened with the statement, "Ecological processes are often not explicitly included in fisheries models and managements, even though they have the potential to impact recovery of exploited stocks, surplus production, biomass partitioning, or biological reference points, and can be of the same magnitude as fishery exploitation." Although cormorant control goals and measures were never within the purview of fisheries managers, it is because of fishery management objectives that cormorant controls were initiated in the first place. The author of the report, Jason Link, remarked that omission of ecological factors may be valid under some circumstances but "in many other cases omitting ecological considerations is imprudent." Link states that cormorants have been shown to negatively impact certain fisheries, such as the lake trout in Lake Ontario, but related how, when prey becomes scarce, as the lake trout did, predators switch to other species, as did the cormorants. He quotes studies showing that "there is little evidence that predation causes large, persistent stock declines," adding that predation has its greatest effects on fish populations that are already depressed. If overfishing, pollution, and invasive species had not initially weakened sport fish stocks, cormorants would have had a less dramatic effect on Great Lakes fisheries.

The USFWS, therefore, was prepared to authorize the killing of perhaps a quarter of a million internationally and federally protected wild native birds per year on the basis of incidental, anecdotal perceptions, the exclusion of ecological processes, without an accurate method for measuring success, and without establishing criteria designed to determine when to stop the killing, if ever. Despite the percentage of overwhelmingly negative comments about the USFWS proposal, in 2003 the USDA/APHIS/WS agency complex accepted the FEIS, but instead of alternative D, chosen by the USFWS, it selected alternative E, emphasizing regional population reduction and increased cormorant killing opportunities. This option permitted the use of lethal measures anywhere cormorants might be found. Nonlethal techniques would *not* be considered a mandatory first step before implementing strictly lethal techniques.

In spite of federal agency rationalizations, the idea of open-ended depredation orders aimed chiefly at one species, the double-crested cormorant, went too far—on legal and humane principles—for many US conservation groups. In February of 2004 a coalition of four conservation and animal-protection groups, aligned with several private citizens, filed suit in the US District Court for the Southern District of New York challenging the government's plan to kill tens of thousands of double-crested cormorants. The Animal Rights Foundation of Florida, Defenders of Wildlife, the Fund for Animals, and the Humane Society of the United States (HSUS) contended that the government's plan violated the MBTA, NEPA, and the Endangered Species Act. Claims that the plan reflected a "knee jerk" reaction with no scientific basis were accompanied by statements that cormorants were beneficial to the ecosystem in that they devoured slower-moving diseased and dying fish, keeping fish stocks healthier. The coalition also referred to numerous studies in which cormorant diets were found to include only minimal numbers of sport fish and had not damaged commercial and sport fisheries.

The stand of the Humane Society took two approaches in defense of the challenge. The first, a "natural" argument, stated, "Whatever impacts cormorants have on fish (or on other birds or vegetation), they have had the same impacts for millennia. The government is trying to control against normal, natural, ecosystem interactions."[17] The claim was that the government, in ignoring undeniable fact, was attempting to regulate an ageless system of natural processes through a series of arbitrary rulings.

The Humane Society's second argument was made on a strictly legal basis. Aside from directly violating the MBTA, it claimed that the USFWS and USDA promoted lethal controls of cormorants without justification, oversight, or management within their own ranks. It added that the agencies had failed to address or acknowledge the value of comments submitted by interested parties. In other statements the HSUS recommended, "Instead of scapegoating cormorants, the US-FWS should address the human causes of declining wild-fish populations (such as overfishing, pollution, and competition with stocked non-native fish)." It concluded, "Killing cormorants is, at best, a short-term fix to a problem that requires a long-term solution."[18]

In the end, conservation efforts failed. The National Sea Grant Law Center reported, "On March 28, 2005, the U.S. District Court for the Southern District of New York upheld the Fish and Wildlife Service's 'Public Resource Depredation Order' to manage the double-crested cormorant population."[19] Among its other analytical points, the court rejected the plaintiffs' claims and found that there was no conflict between the depredation order and the language or intent of the MBTA. The court also ruled that the depredation order was not arbitrary and capricious since conflicting scientific opinions existed concerning damage done by cormorants to public resources, namely, fishery stocks. All other arguments presented by the four conservation groups were dismissed, and the court found that the depredation order was an appropriate way to control the growth of double-crested cormorants. And that was that.

Even though the federal court system had established the appropriateness of the two agencies' actions, the mere issuance of depredation orders and permits, from either state or federal authorities, was not to be the total solution to cormorant conflicts. Writing the plan and implementing it were two different things. Four scientists from the Department of Agriculture's APHIS/WS published a article in 2003 titled "Challenges of Implementing the Double-Crested Cormorant Environmental Impact Statement."[20] Citing at least nine studies, they reported, "Little conclusive scientific evidence has been readily available with regards to impacts of double-crested cormorants (DCCO) on recreational fisheries as well as on habitat degradation on traditional nesting and roosting areas." Again, at the time of the EIS, little was known of the impact of cormorants on free-ranging fish stocks since most of the studies done on fishery damage were related to southern aquaculture sites, not open-water situations. Cormorant flocks feeding on schools of wild fish were typically reported anecdotally by fishermen without tallies of how much actual damage the birds did.

The conference report, in attempting to put the EIS in perspective, raised the issue of alternative E, increased regional population control.

In order for an agency to control the cormorant population on a regional level, it must first determine the current population and determine a population goal that is "both biologically and socially acceptable," with economic and aesthetic concerns also to be monitored. The report deemed the project a "monumental task." For the regional shooting campaigns to be at all effective, monitoring populations before, during, and after implementation would be necessary. The states and tribes involved would have to undertake massive population surveys every three to five years to determine if the culls had had any significant impact on cormorant population growth rates and demographics. After all, these birds, besides being prolific breeders, were aggressively opportunistic feeders capable of infiltrating the ranks of nearby thinned-out flocks, silently replenishing culled individuals. Also, cormorants were extremely mobile, able to leave areas where they are harassed or hunted, only to return when conditions improve.

The lack of scientific underpinning and population data and goals had little or no effect on the depredation ruling. Neither did what the USFWS itself called the "myriad" of other factors beyond the scope of its EIS, which, according to the American Bird Conservancy, includes the impacts of "water pollution, dredging, non-native species, unsustainable commercial fish take, development, erosion, loss of wetlands, climate change, and other factors, all of which play a role in the decline of fish populations."[21] None of these other threats to commercial and recreational fishing interests was considered as an alternative to the obvious cure of killing ever-visible wild cormorants. And so, in disregard of noncormorant causes and the undetermined expenses and results of the FEIS, the APHIS/WS final report concluded with a full endorsement of alternative E, regional population reduction, and its increased sanctioned killing of double-crested cormorants.

The double-crested cormorant was at one time considered extirpated, nearly extinct, throughout its historic range due to DDT and other contaminants. The species then made a miraculous comeback after the banning of the persistent insecticide, and through its own dynamic constitution, returned to, recolonized, and expanded its former breeding range. The cormorant's own increasing numbers and resourcefulness created perceived threats to fishing in the minds of stakeholders and thus put the birds in a government-administered vise and pushed them farther into the realm of being a symbol of ecological conflict.

PART 4

The Channel Cat Comes of Age

12

Catfish on a Shoestring

A Primer

THE TALE OF AMERICA'S catfish industry is the story of a southern industry begun literally out of many a row-crop farmer's desperation to keep his ancestral farmland in his family's hands. For some it was an out and out attempt to escape poverty. For others, owners of extensive farm holdings, the gradual or partial conversion to aquaculture operations was done in hopes of countering slumps in southern row-crop prices. The dawn of America's catfish industry in the United States brought many changes in the South's economy and at the same time created any number of conservation issues, some of which had never before been experienced in the South.

It is legitimate for readers to question why a book about double-crested cormorants would include such a detailed account of the beginnings and development of southern aquaculture. The answer is that close to two hundred thousand acres of exposed catfish ponds lay waiting for tens, perhaps hundreds, of thousands of hungry cormorants to arrive every fall and spend the winter feasting on cultured southern catfish. The tale of the South's catfish industry, a true American success story, runs parallel to the chronicle of the South's conflict with the double-crested cormorant. In order to understand the latter, it's important to understand the former.

To consumers, businessmen, students, educators, and other outsiders living in the Northeast and elsewhere, catfish are not considered much more than weird, whiskered little "slimers" creeping along in the mud holes we drive past every day on the way to work or headed to our favorite trout stream. Not so in the South, where large, meaty catfish flourished in warm streams in such numbers that generations of commercial fishermen made substantial livings supplying them to local markets. Catfish were so sought after by recreational fishermen, and channel cats filled such a large dietary niche, that farmers often engaged in fee-fishing home businesses in which, for a price, they provided access to catfish streams and ponds within their property lines. And in time, as the rearing of catfish grew as a business, catfish promotion campaigns proudly boasted that it could be served in thousands of ways, from the South's traditional fried catfish to trendy almond-crusted catfish fillets sold in northern markets.

A survey of regional and state catfish consumption in the United States conducted for the Catfish Institute in 1998 illustrates the point. The South Central and South Atlantic regions together consumed 177 million of the 281 million pounds of catfish eaten in America that year; just short of 63 percent. The state of Texas alone cooked up 58 million pounds. The entire Northeast region, with an insignificant annual pounds-per-capita figure of only 0.17, in millions of pounds, ate just 9 million pounds. The absolute numbers have surely changed since then, with catfish featured on more upscale northern restaurant menus, as well as appearing on family shopping lists, but regardless of how it's prepared or presented, today the catfish is still not a dinner staple or a generally well-understood industry outside the South.

Aquaculture, a subclassification of agriculture, in that the US Department of Agriculture claims it as within its domain, is defined as "the breeding, rearing, and harvesting of plants and animals in all types of water environments, including ponds, rivers, lakes and the ocean."[1] The "crops" we see being raised at aquaculture facilities vary with regional climates, customs, and tastes and of course differ from freshwater inland sites to coastal and estuary environments. The long and expanding list of critters includes eels, mussels, soft-shell crabs, crawfish, decorative water plants, oysters, trout, tilapia, bluegills, baitfish, ornamental pond fish, turtles for pets, striped bass, and of course catfish.

Fish cultivation was not at all unique to the Americas, and southern farmers were not the first entrepreneurs to explore the propagation of food fish for profit. Historical accounts relate how another warm-water fish species, the carp, was cultivated by the Chinese perhaps thirteen centuries ago and possibly by the Romans stretching back four thousand years. In fact, in the United States, President Ulysses S. Grant directed the newly created US Fish Commission (later to become the US-FWS) to find new food fishes to feed populations of Americans concentrated in the country's expanding cities given the depleted fish stocks in areas such as the Great Lakes. The commission sought new, nonnative species, such as different strains of the Asian carp, which could be cultivated and act as the new protein source for hungry cities. Commission workers, in 1877, introduced more than three hundred carp of various species into a string of freshwater ponds near the city of Baltimore, Maryland, and subsequent generations of the imported carp contributed transplants later released by many state fishery departments into lakes and rivers across the country.

It was hoped that these plant-eating imports would add high-quality protein to the tables of a hungry nation, as well as eat and remove the unwanted vegetation and pond scum clogging many waterways at the time. Appearing on restaurant menus and in homes, carp functioned as suitable food through the close of World War II. After the war, however, more popular species marketed by the growing commercial saltwater industry showed up in US stores and eateries, quickly replacing carp in the American diet. By that time, carp species had settled into lakes, rivers, and ponds by the millions, and had managed to destroy a large percentage of valued aquatic vegetation, and in the process muddied many once-clear waters by rooting into the bottoms of lakes and rivers in search of food.

Other fish that appear in written records include Egyptian tilapia, Roman mullet, and cold-water trout in Europe. Across the Atlantic's waters in North America, the practice of fish cultivation would not emerge until the 1800s. Before that time the natural abundance of wild fish thriving in the continent's lakes and rivers easily met the consumer's taste for fish and there was no need for fish farming. The first attempts at fish breeding in the United States occurred at hatchery-type facilities that were expressly designed to bolster commercial fishing efforts with stockings of public waters.

The next century saw government scientists studying fish behavior and propagation and developing methods to culture and raise commercially valuable species. By 1917, the Kansas Department of Game and

Fish had printed manuals detailing pond construction and fish-farming techniques. And what was by then the US Bureau of Fisheries in Iowa had succeeded in producing viable fingerlings from captive channel catfish adults. Soon after, new federal and state agencies saw the great potential of fish farming and sought to expand studies and disseminate their findings. By the early 1960s a few ponds in Mississippi and other states had produced the first marketable crop of commercially cultured channel catfish. From that point on, catfish production grew steadily, and from the mid-1970s through the 1980s catfish-pond acreage expanded almost tenfold. The catfish industry faced some harder times in the 1990s due to lower market prices but surged again in 1996 when acreage dedicated to fish farming jumped by 4 percent in Mississippi and Arkansas.

Not all attempts to cultivate food fish in the United States were successful. In the 1950s, for example, Arkansas farmers began raising buffalo fish in their already flooded rice fields. Buffalo fish are members of the sucker family and greatly resemble, and are often misidentified as, Asian carp. The buffalo fish gets its name from its humped back, recalling the image of the American plains buffalo and its massive hump. It is common in the freshwater of the Mississippi Valley with some sixty-five-pound specimens reaching four feet. The buffalo fish farming businesses ultimately failed, however, because the rice farmers never determined if there was indeed a consumer demand for farm-raised buffalo fish. There wasn't. The aquaculture industry also faced other false starts in which reserve capital was insufficient to carry novice fish farmers through times of low demand and product surpluses.

The South's row-crop farmers confronted the same issues of cost, supply, and demand that faced buffalo fish producers and probably every farmer in the world. Agriculture in the United States has a long history of dependence on a narrow group of crops. In the South, farmers concentrating their efforts on raising cotton, soybeans, rice, and corn often faced financial disaster when commodity prices fell because their profitability was closely tied to so few crops. Attempts at increasing yields typically worsened the situation in that they reduced production costs only to create larger surpluses, driving prices still lower. Continuing unstable farm incomes compelled southern row farmers to look for new crops. Diversification of crops could hedge their losses, but the introduction of new, alternative crops carried high risks.

The conversion from row crops to catfish farming was slow; initial investments were small and gradual. Two areas in the South where catfish farming found roots and eventually grew to sustainable levels were Mississippi's delta region and Alabama. Eventually, Arkansas and Louisiana joined the other two states to become the four major catfish-producing states in the country. Of the 597 million pounds of catfish the United States produced in 2001, 95 percent of the crop was produced in these four south-central states. The government called catfish farming the fastest-growing segment of US agriculture. In the beginning of the twenty-first century, US catfish-farming revenues amounted to an astounding 40 percent of the world's total fish production.

The South's catfish industry was very much like any other type of farming in that producers needed a specific crop to cultivate. To southern fish farmers that meant just one species, *Ictalurus punctatus,* the channel catfish. Depending on variations in classification schemes, taxonomists describe about thirty-nine species of catfish found in North America. Other than the channel cat, six other species have had their turns at commercial cultivation, but the channel catfish proved itself the leader in the commercial market. Originally *I. punctatus* was found in the Gulf states and parts of the Mississippi Valley but was transplanted and introduced throughout the United States and around the world. The channel cat fits the traditional catfish model with sensitive barbels, or "whiskers," around its mouth and sharp spines in its dorsal and pectoral fins set to inflict a nasty wound when not handled carefully. In its natural environment, which includes fast- and slow-running streams, rivers, lakes, reservoirs, and ponds, adult channel catfish feeding on the bottom dine on a varied diet of insects, snails, crawfish, worms, algae, seeds, and small fish.

Of the thirty-nine catfish species available to farmers, the two characteristics that made the channel cat the most suitable candidate for commercial investment and development were its hardiness and its superior growth rate. This catfish's hardiness is exemplified by the variety of conditions and habitats that suit its lifestyle. Although this catfish prefers clear freshwater environments, it tolerates muddy and even brackish water very well. The channel cat, like all other animals, requires oxygen and favors concentrations higher than 4 ppm but can survive much lower levels of just above 1 ppm. The channel catfish's acceptance of a wide range of environmental circumstances gives the farmer better odds of profitable harvests and a greater chance of overcoming unpredictable climate changes such as temperature spikes and unforeseen droughts.

Just as in row-crop farming, the faster the crop grows the faster it gets to market and the sooner producers turn investment into profit. A fish's growth rate is determined very much by the amount and quality of its food, the quality of the water in which it swims, and the temperature of its environment. Fish generally grow faster in warm water than in cold. In fact, for every 18°F change in its surrounding temperature there is a halving or doubling of its metabolism, directly affecting its growth, or lack of it. The channel cat thrives and grows best in warm water. It experiences maximum growth at the fairly high temperature of about 85°F, an easily reached temperature in shallow southern catfish ponds, and a temperature lethal for many other fish species. At these pond temperatures farm-raised channel cats usually reach a harvestable weight of 1¼ pounds in about eighteen months, a profitable turnaround for the producer.

Again, for the channel cat and any other fish, the abundance and quality of food are major factors influencing its growth. The failure of a captive animal to eat the food supplied by its keeper is a common cause of the captive's death. Not all animals can adapt and learn to feed in captivity. That's not true of the channel catfish. It's trainable. *Ictalurus punctatus* was never going to balance a ball on its nose like a seal, ask for a cracker like Polly, or retrieve a thrown stick, but it could be conditioned to change some its habits. In the wild it lives and feeds on the bottom, but in captivity it can be taught to rise to the surface and feed on floating commercial feed pellets sprayed across farm ponds. When channel cats feed on the surface their human feeders can observe the fish, making sure the commercial fish food is eaten in quantities sufficient to maximize growth rates. Also, because anatomically it has a large stomach that holds a good quantity of food, the channel cat can be conditioned to feed only once a day, thereby simplifying daily feeding routines and reducing energy and labor costs of the producer.

The channel catfish's appeal as a commercial choice for cultivation was further enhanced by its convenient spawning habits, an important function in building and maintaining the producer's inventory. Temperaturewise, the channel cat spawns in water between 75 and 85°F with the optimum in the center, at 80°F, a typical southern catfish pond temperature. This catfish is also very obliging in that it is a cavity spawner, normally building its nests in holes in undercut stream banks or in and around sunken timber and other structures. Farmers wisely took advantage of this behavior by using spawning containers in their ponds or by building specially designed spawning sheds to encourage reproductive activity.

Aside from wisely choosing the channel cat to cultivate, it's no mystery why catfish farming thrived in the South. Two of the major reasons for its growth in the region were the climate and the existence of wide expanses of already cultivated land. The same long growing season that promoted general agriculture and row crops also supported the wide expansion of aquaculture. These long seasons don't exist in the North; thousands of crowded food fish stocked in shallow ponds would never survive a freezing northern winter. The South's warm climate encouraged year-round growth, maximizing catfish yields for the producers. Also, southern land was ready for catfish farming. Farmers had no need to purchase, clear, and prepare new lands. It was more a matter of conversion and change rather than of starting from scratch. The blueprint was in place, but its execution would still require more than a minimal effort on the part of the pioneer catfish farmers.

The Mississippi Delta is described by geologists as an alluvial plain built of sediment deposits from the cyclical flooding of the Mississippi River and its tributaries. The delta is a broad, flat plain spanning over 6,100 square miles formed by the Mississippi's repeated depositions and erosions over time, creating a heavy, clay-based, "gumbo" type soil with a high water-retention capacity, which made it an excellent building material for pond and levee construction. Also, as a product of its geologic history, the delta fortuitously sports a vast groundwater supply, a necessary ingredient in fish cultivation. Not only was the water plentiful, but the water trapped in the delta's subterranean rock layers was easily accessed for filling the farmers' ponds through the drilling of shallow wells, a common, simple, and relatively inexpensive process.

In scale, Mississippi's catfish operations were much larger than those in other states, such as in its neighbor, Alabama. Mississippi row crops were already raised on large farms of thousands rather than hundreds of acres, so the conversion to fish farming was completed on a similarly large scale. Mississippi's larger farms enabled farmers to take advantage of economies of scale. Farmers making larger bulk purchases normally earn quantity discounts from suppliers, which lower the farmers' per-unit costs of important items like catfish feed and fuel for equipment. Large-scale fish farming operations also have high production yields, allowing farmers to sell entire harvests of catfish to processing plants at one time, minimizing transportation costs and ensuring the freshest quality product.

Alabama's fish-farming industry sprang from different roots. It came to life in the first decade of the 1900s after Alabama's cotton crop was devastated by massive boll weevil infestations. In Alabama, farms were appreciably smaller than in Mississippi. They were primarily family businesses in which family members did much of the work, so when conversion to fish farming took place the resultant fish operations were proportionately smaller than in Mississippi. And unlike many of the larger fish-farming operations in Mississippi, Alabama's smaller farms were unable to make the same large purchases and did not qualify for the same bulk discounts and incentives, so their operating costs were higher. And in the sales end of the business, these small, family operations yielded fewer market-sized catfish, often in quantities too small to interest processing plants. As a result, their revenue came from sales to various local markets or from selling directly to consumers, incurring additional overhead in retail and transportation costs.

Today much of Alabama's catfish farming takes place in West Alabama, a trade area comprised of seven counties located in the state's Black Belt region, a geologic area known for its fertile soils.

In contrast to its rich, dark soil, the Black Belt was, and still is, also known for its persistent poverty. The Economics Research Service of the USDA tells how the term *Black Belt* has been used for more than a hundred years to describe the crescent-shaped area characterized by extreme economic distress that runs through a number of southern states. The large black population there is victimized by "high rates of poverty, unemployment, infant mortality, poor health, and low rates of educational achievement."[2] The pre–Civil War cotton plantations of the area were manned and worked by slaves, and today many of the area's residents are direct descendants of those same slaves, still living in impoverished economic conditions in an area unable to make the transition from a simple agrarian economy to one competing in a global market. The one possession some families now owned was the land they had farmed since they were slaves. Since the original prosperity seen in Mississippi never materialized in the Black Belt, the conversion of even small farms to catfish operations was made more difficult due to the lack of equipment, capital, and other financial resources.

Regardless of where the catfish were raised, one of the first challenges catfish growers faced was in the marketing and advertising of their product. People all across the United States ate fish, but regional tastes and preferences determined their choice of fish. Folks along the coasts were more familiar with saltwater species such as flounder, cod, and swordfish. Inlanders were often more comfortable seeing perch, whitefish, salmon, walleye, and trout on their tables. In the South, the catfish ruled, but in the Northeast and western states catfish fillets were unexpected items on local menus. The catfish industry had an uphill battle on its hands in popularizing the catfish as a consumer's regular choice. The catfish's image of a bottom-feeding fish stirring through muddy river bottoms eating algae, snails, and worms had little appeal over fresh- and saltwater predators chasing and gorging on schools of smaller, fresh forage fish. To meet their marketing, advertising, and image goals, the catfish growers of the four largest catfish-producing states banded together in associations like the Catfish Institute, which developed programs to educate the public about the nutritional and economic value of eating catfish. Other organizations with similar goals popped up as the Alabama Catfish Producers, Catfish Marketing Association, Catfish Farmers of Mississippi, Louisiana Catfish Farmers Association, and Catfish Farmers of America.

Slowly but surely the groups swayed America's opinion about catfish. They did their job, and they did it well. The US consumer accepted the farm-raised channel cat as a healthy, less-expensive alternative to red meat and the more expensive fish, like swordfish and wild salmon. In Alabama, for example, over the last twenty years catfish production has grown by 1,600 percent.[3] In 2005 alone, US catfish growers produced 600 million pounds from just 165,000 water-acres, generating 450 million dollars in annual production value.[4] Catfish farmers worked hard to build such an industry, and, as we will see, they worked just as hard to protect it.

Depending on location and resources, the typical modern catfish farm has on the order of twenty ponds, 3 to 6½ feet deep, each about fifteen acres in surface area, set out in grids. Water supplies are more of a problem in Alabama than in Mississippi, since Alabama's rich soil is underlain by layers of natural chalk, which is impenetrable to groundwater. Therefore, unlike in Mississippi, inexpensive shallow wells are of no use. Instead, catfish farms in Alabama depend on surface water derived directly from unreliable rainfalls and runoffs or on more expensive, deep artesian, naturally pressurized wells.

And just as row-crop farmers schedule their plantings and harvests, catfish producers plan their actions as well. The ponds are not stocked and harvested at random. The growers use two different methods of catfish production and propagation. The first, the single-batch cropping system, is the simpler of the two. In this system one group of fingerlings is stocked in each pond, grown to a marketable size of about 1¼ to 1½ pounds or sometimes larger, and then all fish in the pond are harvested together at one time. The multiple-crop method, used more commonly, involves the repeated netting of the pond three to six times per year with a net mesh that only removes market-sized fish, leaving the smaller fish in the pond to continue growing. The harvested fish are then replaced by a new group of fingerlings or larger "stockers." This method has advantages over the single-batch system since it allows the year-round shipping of fish to the processors and markets, eliminating seasonal breaks in cash flow, and multiple-crop harvesting does away with the problems and expense of draining the pond at harvest time.

Working backward in time from harvest to stocking, farmers need to consider what goes into their ponds in the first place. Taking into account losses along the way, the ponds are stocked at extremely high densities ranging from 100,000 to 250,000 fish per acre for fry and fingerlings and 5,000 to 10,000 per acre in grow-out ponds where fish are raised from fingerlings to final harvest. These high densities, if not managed properly, create vulnerabilities of the fish to parasites, disease, and of course hungry predatory birds, including wintering double-crested cormorants.

The conversion of conventional farmland to catfish-farm acreage required more than the simple task of gouging out holes with bulldozers. Success of the conversion depended on the proper design, construction, and placement of the ponds. The USDA supplied guidelines to farmers and conducted site evaluations to help avoid common mistakes from developing into obstacles later in the process. Careless pond and levee construction and planning could result in ponds without practical working access or water supplies, levees prone to erosion, ponds incapable of holding water for long periods, ponds contaminated by pesticide residue, levees unsuitable for bird control activities, or ponds with continuing weed problems.

In addition to all the other elements in planning, ponds needed to be kept to manageable sizes, eight to fifteen acres, to facilitate profitable feeding and harvesting. And as a matter of further cost control and land management, farmers also had to consider the planned use of the pond being constructed, since fry and fingerling ponds were usually smaller than ponds designed for raising market-sized food fish. The conversion from row crops and subsequent pond construction would clearly involve more than just the farmer himself. He required plans and expertise, tools and equipment, additional hands, feed suppliers and processors, fingerlings with which to stock the ponds, capital and loans, and perhaps general community involvement.

Getting the catfish into the ponds was an important first step, but one of the keys to raising any animal on any farm is the type and quality of their natural or prepared feed. For catfish the only manageable feed was a pellet-type preparation specifically formulated for channel catfish. Many feed companies and suppliers developed generic catfish feeds, but these products proved to be inadequate because the companies had no real concept of the nutritional needs of catfish. Irregular deliveries, onsite storage, and transportation costs created additional problems for the growers. A solution was found in the building of local feed mills, owned and operated by associations of neighboring fish farmers. Their proximity assured reliable deliveries and reduced or eliminated storage and transportation costs. These mills recognized and understood the needs of their constituent growers and produced the high-quality, specialized feed farmers needed to turn out superior catfish products.

Animals by and large eat a fairly narrow range of foods. Regardless of what it eats, the animal's food supplies all the nutrients, minerals, and proteins the animal is going to get. For captive animals, their keepers must control and regulate what is eaten and be certain the food presented to the animals contains every necessary element of a complete diet. This is no less important in rearing catfish.

In addition to the content, formulation, shape, and size of the feed, bottom-dwelling, out-of-sight catfish must be observed as they rise to the surface during feeding to monitor their general health, acceptance of the feed, and other aspects of feeding. If catfish don't eat, they don't grow, and they don't make it to market. This means that feedings could not be done by just any field hand, but required an observant, trained fish culturist. Any changes in feeding behaviors had to be noted as they could easily reflect serious problems of contamination or disease, greatly reducing the producer's yields and profits if not properly addressed.

Feeding at the ponds poses other problems as well. Catfish are fed on a once-a-day basis, but how much food do they get? How long do catfish feed? How much does each fish eat? How many fish are there in the ponds? How big are they and how much do they weigh? Do fingerlings eat the same food as fish ready for harvest? In winter, with its lower temperatures, do fish feed less? The answers to all these questions and others are important because too little feed produces smaller fish and the remains of too much feed foul the pond by generating excessive bacterial and algal growth. To answer these questions and avoid other feeding-related problems, some catfish farmers have come to depend on computer programs and standardized models to calculate feeding rates based on a ratio of feed to the total biomass in a pond, but smaller operations no doubt still depend on the producer's instinct and experience.

It is not difficult for the outsider to think of catfish farming as a very low-tech business, and to a large degree it is. But even in the simple everyday task of feeding, there is more to it than meets the eye. Any mistake is costly. Less guesswork results in fewer mistakes; the fewer mistakes, the fewer wasted man-hours and capital. Like most businesses, farm managers and owners who control costs and minimize losses are the ones who turn marginal revenues into acceptable sales numbers and returns on investment.

When catfish farmers look to sales and profits, particularly when dealing with processing plants, they cope with quantity and quality concerns. Every member of the same species is not identical; even offspring born of the same parents differ. Not every farm-raised catfish tastes the same; even individual fish from the same pond may have different flavor characteristics. Some of these variances in taste are undesirable and very noticeable to consumers, who look for consistency in what they purchase, particularly if catfish is a relatively new item on their dinner tables. The industry refers to these taste issues as "off-flavor" problems, and processors never pay top dollar prices for off-flavor products.

Flavor problems that most affect consumers stem from postharvest complications, whereas preharvest off-flavor fish usually never reach the consumer due to taste screenings conducted by processors before they purchase a particular harvest. Postharvest problems as a rule arise from improper handling, causing bacterial spoilage or a rancid taste in fish with a higher fat content. Farmers can easily see their profits dwindle with postharvest flavor questions because there is little they can do to correct them at that point.

Catfish farmers who faced and overcame these obstacles still needed to tackle a concern confronting farmers of all types: disease. Disease is the number-one killer of farm-raised catfish. Catfish are no different than any other fish in that they are constantly bombarded by attacks from other life forms. Besides predators, fish fall prey to disease-causing organisms, and when this happens to captive fish within the confines of a manmade pond the effects can be ruinous to the farmer. These fish-attacking disease agents come in many forms, including bacteria, viruses, fungi, protozoa, and multicellular external and internal parasites. Each one has the potential to weaken the stock, making it less desirable in the market.

Like the diseases themselves, the cures for catfish maladies are not easily managed. Applications of various chemicals and pharmaceuticals may eliminate a particular disease, but drugs not applied in correct dosages by technically proficient personnel may cause additional damage that could require further treatments. Farmers must also keep in mind the federal and state regulations that come into play regarding additives and chemicals applied to food animals. Residues from the applications may also show up in the flesh of the fish or in the effluent of the ponds, raising other issues and tripping environmental alarms.

Prevention of disease may be the best cure of all. Prudent catfish producers can try to avert and control disease-causing conditions in the first place by carefully monitoring the overall environmental quality of their ponds. Growers also contemplate the development of more disease-resistant genetic strains of channel catfish possibly with shorter growing times, greater tolerance to lower oxygen levels, and less affected by the stress created by the harvesting of larger fish in the ponds.

Aside from the costs of construction, maintenance, labor, feed, and fingerlings and problems of off-flavors and disease, catfish farmers also face the second-greatest threat to their investment after disease: raids on their ponds by hungry, fish-eating, predatory birds, particularly *Phalacrocorax auritus,* double-crested cormorants. The birds arrive at their wintering grounds in the delta in November and December and quickly establish night roosts in isolated wetlands or tupelo gum and cypress breaks over water or river oxbows, usually within ten miles of the commercial catfish ponds.

Since the implementation of the 1998 standing depredation order issued by the USFWS, catfish farmers have been permitted to shoot cormorants without permits to protect their ponds but only at the facilities themselves, not at distant roost sites. Shooting on its own had limited results and showed no appreciable decrease in cormorant predation since killed birds were soon replaced by other migrating cormorants. To improve cormorant control, shooting at the facilities was combined with many nonlethal harassment techniques in what was considered an integrated control plan.

Even in the face of on-site shooting programs cormorants gathering in night roosts continued to plague the producers when the birds took to the skies each morning and headed for the ponds. In Mississippi in 1993, USDA's Wildlife Services instituted a regional roost dispersal program aimed at cormorants in which multiple roost sites were harassed simultaneously using noisy pyrotechnics. These "major pushes" worked as immediate solutions but for the most part produced a redistribution effect rather than an elimination of the birds. Eventually the birds settled back into the roosts.[5]

In Alabama, farmers and wildlife biologists wage a virtually continuous war against wintering cormorants. The biologists and aquaculture specialists of state and federal agencies in catfish country work closely with their clients, the catfish producers, with rarely a day off. Jerry Feist, a wildlife biologist for the USDA's Wildlife Services who grew up in North Dakota, has an office wall covered with local maps dealing with current cormorant locations. Pins mark the known twenty or so neighboring cormorant night roosts. In speaking with the author, Feist said the general plan was to "bust them out in the evening," meaning chasing the cormorants from roosts, shooting some in the process, and attempting to keep them moving away from the ponds. And it's not always an easy task since many of the roosts are isolated and only accessible by boat.[6]

Much of the information concerning current cormorant whereabouts is gathered by the producers themselves. In seminars, Feist tells the growers two things when they call in to report flocks of cormorants hitting their ponds. First, "get 'em gone," clear the birds off the pond, and second, make a note of which direction the cormorants are coming from so WS can locate that particular roost. One catfish producer related how at first light, when the cormorants leave their roosts, he looks for the flight lines of birds preparing to raid his ponds. The producer and his staff wave, shout, and shoot to keep the birds moving on to some other producer's ponds, where they will be chased off in turn. The birds continue to move until they find a careless farmer's unguarded ponds to settle on and feed. In addition, the producer uses propane cannons and other noisemakers, as well as the deployment of old cars and trucks, which he has towed to different locations to discourage cormorants from landing. Growers in the area generally face a 55 to 65 percent survival rate for farm-raised catfish, with about half of the loss due to wintering cormorants.

Greg Whitis, an aquaculture extension specialist at the Alabama Fish Farming Center in Greensboro, who works closely with Jerry Feist, said that 99 percent of catfish farms in the state are family operations and that margins for producers are so slim that controlling cormorant predation is the key to turning a profit or perhaps going under. Whitis agrees that "roost harassment" is the primary method of cormorant control. The process is simple in design: the double-crested cormorants are spotted during the day "by planes or ground intelligence" and are then chased to new locations later that day. Citing the fact that "there is no pristine habitat to move them to" and that they'll only settle and roost near other ponds, the only answer is to "keep them moved."[7]

Cormorant numbers do fluctuate a great deal in Alabama's catfish country. The midwinter survey counts cormorants in and around catfish country, as well as other concentrations that might pop up, but the survey numbers reflect essentially only those birds with a potential to raid the catfish ponds. The wintering cormorant population peaked in 2004 at 36,000, dropped to 26,000 in 2005, ranged in the low to midtwenties through 2009, and reached 27,000 in 2010.[8]

The reports generating statistics of predation by cormorants tell a sad story for the catfish growers. According to Wildlife Services, "[M]ore than half of all farmers and ranchers experience some kind of wildlife damage each year."[9] When it comes to catfish farms, the majority of catfish operations in the four-state area of Mississippi, Alabama, Arkansas, and Louisiana are affected by avian predators. A 1996 survey of catfish producers revealed that 77 percent of operations in Mississippi, 66 percent in Arkansas, and 50 percent in Alabama claimed losses due to cormorants, and with cormorant numbers on the rise losses were sure to increase. Wildlife Services estimates that the growing cormorant population consumes 18 to 20 million catfish fingerlings each year, producing a total annual financial loss of about 25 million dollars. A three-year study conducted in the Mississippi Delta region found that 50 percent of the diet of cormorants wintering there was made up of farm-raised catfish. Some studies show that cormorants feeding at a pond are capable of consuming an average of 5 catfish per cormorant per hour, with rates as high as 28 for cormorants working fingerling ponds. With just 30 cormorants feeding at a 20-acre pond stocked with fingerlings, a density of about 125,000 per acre, the fish population would be halved in just 30 days. A single cormorant consuming only catfish would destroy about a dollar's worth of fingerlings every 24 hours. Other scientific studies show that as cormorants feed and pursue catfish in the ponds they also leave in their wake unknown numbers of fish injured in the bird's pursuit and damaged levee walls, and their very presence in the ponds disrupts critical feeding schedules. Similar reports and articles abound in periodicals such as the *Farm Press,* the National Aquaculture Association's website, and the *Journal of Fisheries Management and Ecology,* as well as in several university extension publications. These circumstances of course represent "ideal" conditions for the double-crested cormorant, but any scenario even close to one of these denotes financial nightmares for southern catfish growers.

Just as cormorants evolved and adapted to environmental changes through the ages, modern cormorants adopted learned behaviors that improved their hunting success at catfish ponds across the South. Describing cormorants preparing to fish the ponds, some catfish growers referred to the birds, with an accuracy derived from experience, as "thieving sentries." The birds' fishing activities are at times so intense that all available hands are called away from their regular duties to deal with raiding flocks of up to five hundred birds, reducing productivity and inevitably sales and profit. Some farmers observed that large groups of double-crested cormorants worked in shifts and exhibited cooperative teamwork in which the birds herded channel cats into one corner of the pond where they took turns, alternating standing as a curtain to first block escapees and then as hunters feeding on the trapped fish.

As these farmers learned, double-cresteds are very intelligent, instinctive, and adaptive birds. The same attributes that enabled the species to survive near extinction made the cormorant a tough adversary, setting hurdles for catfish growers in the development and implementation of protective options for their ponds. Cormorants are also known for their persistence and perseverance. In their attempts to manage and control cormorant populations, growers, along with local, state, and federal officials, before very long discovered the bird's cleverness and sought to stay one step ahead of *Phalacrocorax auritus* in both their lethal and nonlethal containment programs. It made for a shifting and challenging game.

Catfish growers on entering the business confront so many obstacles that it's easy to think that the odds of success are stacked against them. Of all their losses, kills due to raiding cormorants rank second, outdone only by disease. With so many drains on their revenue and profit, it's not unreasonable to ask why these farmers and families do it at all. Why not just work in the factory, the office, or the store? In her 2006 book *Fishing for Gold: The Story of Alabama's Catfish Industry,* Karni Perez looked at this issue.[10] She pointed out that many farmers in Alabama wanted to carry on their long-held rural traditions and remain on land that had been in the family for perhaps generations. To them farming was not just about returns on investment; it was their way of life. As Perez put it, catfish farming was different than other aquaculture businesses in which "most investors come from the outside and want to 'make a killing' at them. Alabama's catfish farmers want to make a living at it." Perez explains also that these farmers developed a "feeling for the natural conditions" of the land they lived on with a fascination for the ponds, the water, and the fish.

Perez and other writers, in print and online, discuss how the close-knit, small-scale farming operations brought communities together, especially in Alabama. Since all the catfish growers at one time or another would encounter the same or similar production snags, one farmer's problem became every farmer's problem. Communities frequently banded together to search for common solutions and also in resolving conflicts arising between farmers.

When Americans consider which enterprises US ingenuity created, in addition to baseball, basketball, the music of jazz, rock, and country, crazes such as the hula hoop, and the massive industries once revolving around Detroit's automobiles, catfish farming falls fairly far down the list. But the truth is that America's catfish farming began in the minds of a few men forced and willing to use and develop what local natural resources were available to them. In the four southern states billed as major players in the industry, Mississippi, Alabama, Arkansas, and Louisiana, that resource was the channel catfish.

And at the same time that the channel cat's popularity as a crop and food fish expanded, the double-crested cormorant rebounded from its near extinction due to pesticides and pollution. In a matter of a decade or two the cormorant replenished its numbers to near-historic highs, creating the unanticipated situation for the growers of too many cormorants with too few apparent options. The expenditures of producing and harvesting farm-raised catfish filled a number on lines on ledger sheets including feed, wages, interest, costs of fingerlings and other fish stocks, fuels, transport, and certainly bird management. Catfish farmers had a lot invested in their businesses. So, even though double-cresteds were deemed protected under federal laws and international treaties, the South's catfish farmers felt they had to act to save their regional, hometown, family businesses and the communities that depended on them for their existence. And, unfortunately for the cormorant, that action meant dangerous, bloody encounters with men with firearms. These same defensive actions catfish farmers thought were necessary also brought about fierce debates and conflicts with conservationists, naturalists, legislators, and others, who may have been less aware of what was at stake for these growers: their long-held, cherished farming traditions and their economically important catfish industry. The growers in turn may have underestimated the value cormorants had in circles outside the aquaculture community.

13

The Annual Battle of the Ponds

THE *New York Times* archive contains an article by Ronald Smothers from 1989 titled "It's Fish Farmers vs. Cormorants, and the Birds Are Winning."[1] The piece described how double-crested cormorants once flew over the vast "white gold" cotton fields of Mississippi's delta region on the way to their wintering grounds on the Gulf Coast, and now, instead of overflying rows of brushy cotton, the birds spy broad grids of slick, shallow, fifteen-acre ponds stocked levee-to-levee with tasty farm-raised channel catfish in densities as high as sixty thousand catfish per acre. The delta has become a destination for migrating cormorants rather than acting as a guidepost directing them southward to the Gulf as it did in the past. Smothers then compared the catfish farm to World War II London during the blitz, as propane cannons fire sonic blasts instead of artillery shells into the sky to frighten off circling cormorant "bomber" formations looking to attack the ponds below. Smothers created a great mind's-eye battle image. Thousands upon thousands of double-crested cormorants, from breeding grounds as sizable as eighteen hundred miles long, "funnel into the heart of catfish farming country, at the confluence of the Arkansas and Mississippi Rivers" and set the stage for discussing the conflict between southern catfish farmers and their evil nemesis: the double-crested cormorant.[2]

Catfish dinners are respected table fare in the South. "Mississippi catfish producers want customers to eat lots of their product, but when those consumers are predatory birds, it's time to get out and patrol the ponds."[3] A little humor lightens the situation to a point, but protecting the ponds from the birds is serious business. In 1989, a period between the ratification of the MBTA and the adoption of Depredation Orders in 1998, sonic cannons, other fright tactics such as human "scarers" cruising the levees in pickup trucks, and a few mechanical deterrents were about the only legal weapons catfish growers had available to skirmish with raiding cormorants. The birds were internationally protected and yet hated throughout catfish country. On the delta and in other southern locations, cormorants were seen as not much more than feathered boll weevils, devouring catfish instead of cotton to the estimated annual cost to the growers of twenty-five million dollars in cormorant damage and damage control. Killing cormorants was not yet legal, but more and more, between friends, fewer and fewer people were claiming that it wasn't being done.

The 1998 Standing Depredation Order changed those neighborhood conversations. Growers now kept logs, available for official inspection, documenting the number of cormorants they killed each month. And the recorded numbers were lower than what would be expected to have any real impact on local cormorant populations. One study on cormorant "takes" showed that in one nine-year period, 1987–95, 847 depredation permits were issued to southern aquaculture facilities authorizing the taking of just short of 55,000 cormorants.[4] The *recorded* take was listed as 35,332, only 64 percent of the total number permitted. Over that same nine-year period, then, the average total take per permit was about 42 cormorants. The study also related how cormorant populations remained stable and even increased in some areas. The authors concluded that the tens of thousands of recorded cormorant kills at southeastern aquaculture sites "did not adversely affect regional winter or continental populations of these species." The reason for so small an effect is that the regional winter population of cormorants in the Southeast is measured in hundreds of thousands and the continental population amounts to at least a million, and possibly twice that number. Numbers like 55,000 and 35,000 amount to only a small percentage of the total. In some breeding areas cormorants were expanding their species presence by more than 30 percent annually, far exceeding what the catfish growers shot. So what was all this killing about? And how much of that twenty-five-million-dollar loss did the growers recover?

Aside from probably not being able to physically eradicate cormorants from areas where humans raise and harvest fish, there exists a recognition described in a USDA staff publication that "the double-crested cormorant is a native species that is of intrinsic as well as esthetic value to humans."[5] Depending on where and how they are produced, fish stocks comprise public and private resources; humans laying claim to these stocks openly kill cormorants with little regard for the actual results and effects of the culls. What is really lost in the process, buried in the numbers, and inflammatory rhetoric, is the idea that the birds themselves, the cormorants, are a *public* resource. Double-crested cormorants do come with benefits through psychological, ecological, and economical routes.

Birdwatching, as one route, is often described as the number-two hobby in the United States, second only to gardening. More than forty-fix million people follow the behavior and natural history of birds.[6] Birders use numerous techniques and technologies to record and savor their hobby experiences, including: photography, video and audio recordings, online and print journals, life lists, clubs and associations, competitions, and the Christmas Bird Count. They invest millions of dollars in cameras, binoculars, spotting scopes, books, recordings, workshops, cruises, and tours. In Arizona alone, more than three hundred thousand birders visit the state every year for glimpses of migrating birds, adding about a billion dollars to the state's economy.[7] It is a popular pastime important in many people's lives. And along the coasts, lakes, rivers, and estuaries, cormorants, their active pursuit dives, their grunty calls, and their resilient life histories, fill a niche in that pastime.

Another benefit of supporting cormorant populations is that the sensitivity of cormorants to toxic industrial chemicals and persistent insecticides is well documented in popular and scientific literature. We know that DDT and PCBs nearly exterminated cormorants through disruption of their reproductive cycles. Cormorant populations, then, are living gauges of environmental quality, their expanding or collapsing populations acting as potential indicators of environmental contaminants endangering birds as well as humans. If cormorant populations begin to crash it would be wise to look for the reasons. Healthy cormorants are indicative of a healthy ecosystem. The realization of such a relationship in the public's mind occurred only after the hard-fought battles by conservationists like Rachel Carson.

Double-cresteds are aggressive fish eaters, but their hunts and feeding don't always automatically conflict with people's interests. Sometimes they eat fish species humans wish they would eat. Regardless of how they appear, catfish ponds are not pure catfish culture dishes. Every now and then they are contaminated by unwanted fish species such as gizzard shad and members of the sunfish family. These stray fish can enter pond systems through flooding rivers and creeks, careless pumping errors, and unauthorized dumping of baitfish and panfish into the ponds. These unwanted fish compete with the channel cats for food and may be carriers of various diseases and parasites, which could devastate the ponds. Cormorants love sunfish and gizzard shad and can help rid them from the ponds as well as eliminating slower diseased individuals of all species in the ponds. In the wild, cormorants eat millions of forage fish. And similar to the ponds, the first prey taken are the slower, genetically weaker, and diseased individuals, helping to stabilize the predator-prey balance and reduce the spread of disease in waterways like the Great Lakes, reservoirs, and rivers.

If we accept, at least to a small degree, the premise that cormorants are a useful public resource, then it is logical that wildlife managers should seek out evenhanded approaches to cormorant control in their management philosophies. The APHIS/WS program, for instance, aids traditional farmers and aquaculturists by providing leadership, expertise, training, equipment, and direct intervention in wildlife conflicts. The agency's primary goal is to reduce damage to agriculture and aquaculture, but it must balance its protective actions with respect for the wildlife in question, which in our discussion is the unloved cormorant. The same USDA staff publication cited earlier offers a clarification of its agencies' policies by stating that "[R]esponsible wildlife managers balance the needs of humans and wildlife, foster tolerance toward wildlife, and advocate cost-effective and environmentally acceptable remedial solutions that reduce the implementation of environmentally unacceptable actions by those experiencing the problem."[8] The APHIS/WS staff writers summarize what it takes for agency managers to successfully oversee wildlife populations: "Managing a resource to reduce wildlife conflicts usually involves modifying cultural practices (e.g., animal husbandry or crop selection), altering the habitat to reduce its attractiveness to wildlife, or adjusting human behavior. In the case of aquaculture production, the objective would be to reduce the vulnerability of fish to predation by cormorants."[9] Notice that the only offending wildlife mentioned in the statement is the cormorant; not deer, coyotes, raccoons, crows, geese, or blackbirds.

In a wildlife damage and management situation one path to the solution is through an examination of the cause, and that cause is not always a matter of fact but a function of perception and convenience. One very philosophic paper written by a marine sciences researcher in 1995 described how wildlife control functions in real life: "In wildlife management, only part of the process is scientific."[10] As an explanation, the author, D. C. Duffy, added, "Most problems and solutions involve human attitudes and actions which serve as a filter for the science, so, in dealing with the problems of the Double-crested Cormorant (*Phalacrocorax auritus*) competing with fishermen, we have many scientific facts, but there are differing opinions about the problem and what can be done about it." Duffy admits that science only goes so far but asks nonscientists "to examine what the science has to say with an open mind" because "things are not as they appear or as we may wish them to be when dealing with cormorants." Duffy related his philosophic tenor to fishermen, but his ideas could just as easily be applied to catfish producers and their hard-line attitudes toward cormorants.

Quoted in the *New York Times* by writer Ronald Smothers six years earlier, an animal control officer presaged Duffy's wisdom, saying, "In a sense, man has created this problem through degrading the natural environment by draining and filling in natural feeding areas or contaminating them with years of pesticide use. Then he creates a new prey base for the cormorant, and it is only natural that they turn to it. It's not the cormorants' fault, and shooting them is not the answer."[11]

The MBTA granted protection to virtually every bird not covered by earlier legislation with only a few exceptions, like the starling; the random shooting of thousands and thousands of birds, including double-crested cormorants, came to an end. The general persecution of cormorants had ended, but man's creation of a "new prey base" also shifted cormorants into a new domain. In the words of WS wildlife biologist Jerry Feist, as a result of the cormorant's strong reappearance "the bird was bastardized."[12] Cormorants were now an intrinsic problem, a bird with no inherent value, at least to the South's catfish country. In other areas, however, cormorants are thought to play an important role in maintaining the predator-prey balance in the wild and, through the lenses of binoculars, in soothing human psyches.

Every year cormorants come to the ponds, as they have now for more than three decades, and every year catfish producers confront the birds in the same ways, using the same methods. A written abstract of a presentation to the Alabama Fisheries Association by Jerry Feist concluded "It is not likely the population levels of DCCO will drop dramatically any time soon, so commercial and sport fish managers need to know what tools are available to help manage predation from DCCO."[13]

When asked by the author why catfish producers never changed their plan of action or sought to really improve protection of their ponds, Feist said he wondered about that frequently, almost daily, as he drove through Alabama's catfish country and around its many ponds. Mentally asking himself why the technology never advanced past a certain point, and dreaming up outrageous, impractical control methods of his own, he still has no answers—except to keep doing what he's been doing for his clients—the catfish producers.[14] His thoughts reflect how the industry in both Alabama and Mississippi began as shoestring alternatives to farming row crops and to a certain degree never moved far from that mark, looking to preserve the simplicity, down-to-earth nature, and low-cost operations of raising catfish.

Cormorants are here to stay; they have carved out ecological nesting niches on North American lakes, estuaries, and coasts and found gainful winter employment at southern catfish ponds. And if it's true that humans, not the clever, opportunistic cormorants, are responsible for creating the conflicts through their free offerings to the birds, then humans must be held accountable and responsible for the solutions. And no one ever claimed finding the answers was easy. In the same *Times* article mentioned earlier, a catfish grower admitted his frustration with the nuances of the *Phalacrocorax auritus* conflict, actually crediting the cormorant with a positive attribute, saying, "A good fish farmer can handle things like disease and heat spells by just working harder, but here you're dealing with a thing with a mind and they will run you ragged."[15] Finding solutions takes time, especially when the problem has a "mind" and fights hard for its share of the proceeds. But questions then arise of how long this process will take and how producers will get to the finale.

It's probably unfair to blame the original players in the catfish industry for not foreseeing the future onslaught of cormorants, although other birds, like great blue herons and pelicans, were also quick visitors to the ponds. But what of the later entries into the catfish game? As more row-crop acreage was converted to grids of catfish ponds, their owners and managers either missed or ignored the potential threats of double-cresteds to their businesses. From the early 1970s through the beginning of the twenty-first century little was accomplished by catfish growers in defending their ponds against their feathered enemy "with a mind." A university study on the impact of cormorants on the farm-raised catfish industry in Mississippi showed that in an estimated 2001 *statewide* catfish farm budget of approximately $295,000,000 only $725,000 was allocated to "bird management;"[16] less than the figure for electricity or fuel or chemicals, even less than the budget line for "miscellaneous." As a fraction of the total budget, expenses for bird management measures amounted to about one quarter of 1 percent. Dr. Jim Steeby, an aquaculture extension specialist for Mississippi State University and former Peace Corps volunteer who worked on aquaculture projects in India and Africa, today works directly with Mississippi catfish farmers improving production and worker safety on the farms. Dr. Steeby commented that much of the operating cost of bird management on the farms is spent in increased vehicle costs, fuel, and labor expenditures as workers relentlessly patrol the ponds in search of raiding cormorants. Additional outlays also occur in the repair of the levees from damage caused by farm vehicles using them as causeways to the ponds and from raiding birds in the wet winter months.[17] Considering how economically important the cormorant issue is to the catfish producers, and how the growers have demanded both congressional and state legislative relief, the catfish farmers themselves, by not committing specific funds to cormorant control, give the impression of rating the cormorant conflict low on their list of financial priorities. And yet, as a cause of damage to the catfish stocks, cormorants are ranked second, after disease. In the *Delta Farm Press,* a fish farmer, who in addition to channel catfish also raises carp, crappie, bass, and bluegills, said, "We keep employees running birds every chance we can. We don't have enough employees to dedicate any solely to cormorant control. But when the birds get bad enough we have to stop what we're doing and deal with them."[18]

Although federal officials failed to authenticate exact catfish losses to cormorants, it is well documented that the birds beat the farmers out of tens of millions of catfish fingerlings every winter, amounting to millions of dollars in damages. Unlike the sport and commercial fishermen of the Great Lakes, who claim ownership of a public resource in their fish harvests, southern catfish producers *do* own their fish stocks and are entitled to protect their property. But catfish farmers still seem shocked and outraged that the birds help themselves to free catfish meals. In the forty years since DDT was banned, aside from a few scare tactics, the best remedy the government and the industry could muster was to shoot thousands of cormorants, having little effect either at aquaculture facilities or on the overall cormorant population.

Also, in fairness to the catfish farmers, an alternate way of looking at the problem is to face the reality that increased costs to the farmers result in higher costs to the consumer. Catfish producers and processors combined get about $3.00 out of the $7.00 or $8.00 per pound charged for catfish fillets found today in supermarkets in the Northeast. Catfish is still one of cheapest "seafoods" found in the market. Recent price checks show farm-raised Atlantic salmon at about $10.00 to $11.00 a pound, wild salmon at $16.00 to $18.00, medium shrimp at about $10.00, and swordfish at about $16.00. Even in a depressed economy there is still room in the market for a small boost in catfish prices, although it might have to be accompanied by an increased promotional effort to build more consumer interest in American catfish fillets for the table as opposed to imported products. Recognizing the influence of market pressures and the fact that catfish farmers many times share ownership of the feed and processing plants through their associations, they have some control over the prices offered to supermarket chain buyers. Small negotiated wholesale price increases might well be used to finance the payroll for workers specifically hired to deal with cormorants. Such a plan might reduce the need for culls and shootings at aquaculture sites. Including *all* the costs of getting a product into the hands of buyers and consumers is the only way to develop a realistic budget or business plan. Ignoring actual cormorant control costs inevitably tilts the scale in favor of the birds, since there is really never enough money to spend in solving the problem. It may be far easier to complain, shoot thousands of wild birds, and hope for the best.

One effect the shooting and culling programs *did* have was to further polarize other stakeholders and the public at large. The controversy brought out politicians who made outrageous, unsubstantiated claims about the numbers and weights of fish cormorants could consume in a twenty-four-hour period in an attempt to pass drastic anticormorant legislation. Like the Internet of today, once those types of figures circulate through information networks, they are apt to keep showing up as "fact." The National Aquaculture Association, whose banner reads "One Industry, One Voice," in a long series of position statement "whereas" allegations, followed by several "be it resolved" motions, strongly encouraged congressional action in the implementation of "population reduction measures for DCCs and other depredating birds."[19]

At the other end of the spectrum, conservation groups fought uphill battles defending cormorants from their detractors. They asserted that the cormorant's recovery from near extinction was a near-miraculous event worthy of celebration, along with the ecological comeback of the lakes and wetlands in which they thrive. It is their contention that the cormorant deserved to reach its past historic populations and even exceed them. Cormorant Defenders International (CDI), primarily involved in Great Lakes cormorant issues, describes these birds as being "misunderstood, maligned, and mistreated." The organization argues that the double-crested cormorant controversy was fueled by anglers and wildlife managers in claims that nature is out of balance in "an organized war" waged against cormorants.

The "war" waged at southern aquaculture sites is probably not as "organized" as the CDI and other groups would have the public believe. At the ponds the name of the game for the producers is vigilance. Workers patrol the ponds repeatedly during daylight hours to chase cormorants off the ponds. With no limit on how many birds they can kill at the ponds under updated regulations, Dr. Steeby observed that workers only shoot as many birds as necessary to instill a fear of the patrols in the remaining cormorants. Since there are hundreds of thousands of cormorants roaming the sites in winter they can never hope to make a dent in the population. Dr. Steeby remarked that "cormorants have their place in the environment," but the present-day world is not the world of three hundred years ago, when fewer cities and towns and people and industries existed to generate the thorny conflicts between humans and cormorants we see today.[20] His analysis suggests that there simply may not be room, or sufficient natural support, in the ecosystem for as many free-roaming cormorants as there was centuries ago. Therefore, controlling their numbers to establish a realistic twenty-first-century balance should not be out of the question. But who is wise enough to determine how many cormorants it takes to achieve that balance, and how is that measured?

As cormorants encroached on new territories, the conflict's polarizing effects spread even into the controlling and regulatory agencies themselves. We have already seen how at least one animal control officer thought shooting cormorants was not a valid control measure and that cormorants should probably just be permitted to go their own way. Other officers in other agencies saw things differently.

"Flat out, we need to kill a whole bunch of these birds" is the quoted sentiment of Thurman Booth, who in 2002 was the Arkansas state director of USDA's Wildlife Services.[21] The agency itself has a stormy background, which is reflected in its attitude toward wildlife control. In 1985 Wildlife Services was shifted to the authority of the USDA. Previous to that, the agency was an arm of the USFWS under the Department of the Interior. According to Booth, people in the USFWS "aren't hunters and philosophically disagree with the concept of killing anything." He accused them of staying inside, reading old literature, and having little contact with wildlife. Booth claims that rather than considering the problems catfish farmers face, the USFWS has unhealthy relationships with animal-rights groups such as People for the Ethical Treatment of Animals (PETA) and the Humane Society. As a great champion of the farmer, those in the USDA, though continuing to deal with food and fiber production, at least "understand that in order to have a steak for supper, something has to be killed." It is WS that has direct contact with the farmers, not the USFWS. The department-agency polarity results from the fact that WS has the responsibility of protecting farmers' assets from destructive wildlife but not the authority to act on its own. So when it proposes lethal (or even nonlethal) wildlife control measures they are subject to approval by USFWS officials, the ones Booth claimed are office bound and isolated from true wildlife issues. Booth also postulated that when government agencies fail to respond to the needs of farmers incidents such as repeats of the Little Galloo Island shooting and even general chemical poisonings could become much more commonplace.

Outside the agency versus agency turmoil, the Humane Society also has had a voice in the lethal versus nonlethal controversy. It asserts that USFWS allowed the depredation orders to continue without first forcing producers to exercise "due diligence" in exhausting nonlethal measures before shooting birds, that is, catfish producers should have to prove they did more than nail up an owl replica to claim nonlethal measures were tried and failed. Another criticism levied by the HSUS was that the USFWS never established a baseline (in dollars or percentages) for losses and damages due to cormorants before lethal measures were allowed to be used; loss to predators is a recognized cost of doing business. Instead, it was left to the catfish producers to decide when nonlethal costs got too high. Humane Society spokesperson Stephanie Boyles, speaking of the USDA/WS, also implied that in a sense the agency helps to protect itself and the agency's jobs by ensuring "return business"—they benefit in the long term by not actually solving the problem.[22]

Today's USFWS has fewer links to cormorant control than it did in the past. Senior wildlife biologist Dr. Albert Manville, a fourteen-year veteran of the agency, said that he was not aware of any active participation in cormorant issues at the time, since other issues, such as communication towers and wind power projects interfering with bird migration, overshadow them. In a position not dissimilar to that of Stephanie Boyles, Manville agreed that government USDA/WS personnel were doing much of the work protecting the ponds at taxpayers' expense. He called the process "a one way street"—with WS doing what the producers should be doing themselves.[23]

The South's catfish producers, besides creating a massive food supply, may have also made cormorants healthier. The growth of wintering cormorant populations has been directly linked to the expansion of the catfish industry in the South through a substantial decrease in the bird's mortality rate. Studies show that the "body condition reflects the potential ability of a bird to cope with its present and future needs."[24] The idea that stronger, healthier members of a species have better survival rates than weaker individuals is not a striking concept. The interesting, noteworthy aspect is how the idea applies to cormorants.

An analysis done of the body mass of cormorants collected in the delta region showed that cormorants, primarily adult males, roosting near catfish ponds built up more body fat than cormorants wintering in areas with less catfish production.[25] Fat is a tissue in the body that stores energy; we have all seen and heard how mammals, such as bears, store fat for winter hibernation and how whales build fat reserves for their long migrations, courting, and breeding. In the same way, catfish protein helps build a fat layer in double-crested cormorants for their long migratory flights north in the spring. In addition to cormorants building bulk at the ponds, two other behaviors add to the cormorant's acting as a thorn in the paw of the catfish producer.

On their wintering grounds near the catfish ponds, cormorants undergo a change in their feeding behavior in the months of February and March. The birds enter a stage medical doctors and scientists call hyperphagia, or excessive-eating.[26] As many catfish as producers observe double-cresteds taking at other times, the number is magnified when the birds exhibit hyperphagia. The birds begin moving north in April, so they literally "fatten up" on channel catfish for the exhausting flight ahead of them. The second behavior that contributes to the cormorant's welfare is an action by the producers themselves, for it is in March that farmers stock their ponds with millions of catfish fingerlings to begin the upcoming growing/feeding season. It's perfect timing—for hungry cormorants. The birds need to engulf as much food as possible in the shortest amount of time, and the growers fill their ponds, the birds' dinner bowls, with as many channel cats as they can, as fast as they can, to maximize harvest yields. Therefore, when the birds return to their now traditional northern breeding grounds in April and May, they are as healthy and fit as they were the previous summer, gorging on Great Lakes alewives, perch, and of course those tasty smallmouth bass that so distress the fishermen. The growers are certainly happy to see the cormorants leave, but there has to be an undeniable sadness in that it is calories generated from eating *their* catfish that fuel the bird's northward migration. Thanks to the growers, the birds will arrive at their nesting sites healthier, and will breed sooner and more successfully, only to return again in the fall for those carefully laid out catfish meals.

To defeat this inherent sadness, catfish producers developed a number of tactics to dissuade cormorants from eating their channel catfish crop. Between 1972 and 1998 cormorants were protected from shooting by the MBTA, so the defenses the growers had at hand were limited. A USDA science-based initiative set down the foundation for controlling wildlife conflicts with human interests: "Alleviating wildlife damage entails employing one or a combination of three strategies: 1) managing the resource being protected; 2) physically separating the wildlife from the resource; or 3) managing the wildlife responsible for, or associated with, the damage."[27] As we'll see, none of these lines of attack comes without some sort of trade-off by the catfish grower.

Managing the resource, in our case the catfish ponds, is not a simple matter. Each pond spans about fifteen surface acres, with twenty or more ponds linked by levees. Smaller ponds make for easier installation of control measures, but force farmers to build more levees, thus incurring greater expenses in pond construction and maintenance. And above all, smaller ponds mean smaller yields and reduced profits.

Another way of managing the ponds is to lower the density of stocked fingerlings and harvest-sized fish, making fishing more difficult for the cormorants. This method also negatively affects production, reducing cash flow. Also, stocking fingerlings later in the season, in April rather than March, after many cormorants have already headed north, could reduce the number of fish susceptible to predation when cormorants shift into their overeating mode, but this change also would shorten the warm growing season for the new crop of catfish. Neither of these two approaches, standing alone, resolves the conflict for the growers.

The USDA's alternate concept of separating the wildlife from the resource means erecting physical barriers to exclude the offending wildlife.[28] Cormorants are clever little devils in that they can fly, walk, swim, or dive over, under, or around almost any type of enclosure. But success by the grower is not measured on an all-or-nothing scale. Blocking cormorant access to the ponds by even a respectable percentage could be grounds for success and bragging rights. Virtually every exclusion method, however, poses the problem that it interferes with pond routines such as levee maintenance, feeding, and harvesting. Structures like supported netting, plastic or wire grids above, on, or below the water level, and overhead wires usually have to be removed to service the ponds. Extra effort translates into higher payrolls. And the materials to build the grids and support the nets, and the nets themselves, become expensive when talking about covering farms of 300 acres, and perhaps ten times that number. "Separating birds from catfish is an expensive proposition."[29] An exclusion system might cost more than five times the expense of the original pond construction, again increased expenses for the growers.

Besides changing their practices or building cages around the ponds, the last alternative left to the fish farmers in the USDA's scheme is to manage the wildlife. In the 1980s, the 1990s, and the first years of the twenty-first century, lethal measures were very limited or not permitted at all. What was then left to the producers was their ability to scare and harass the birds. Cormorants are clever, and producers, being clever in their own way, continually came up with ways to counter the birds' raids. Scare or frightening strategies vary with time and the measures to which the cormorants become accustomed. Methods range from plain noisemakers to pyrotechnic devices (projectile firecrackers), screamers, horns, bangers, old cars and trucks indicating a human presence, propane cannons, automatic exploders, scarecrows and simple effigies, evil-eye balloons, flash tape, cormorant distress calls, and automated inflated human figures dressed in garb similar to what live workers wear. Cormorants are sovereigns of opportunism; the trait is deeply implanted in their genetic makeup. They are quite capable of habituating to frightening devices, particularly those that provide no real negative reinforcement. Producers and their employees must, to revisit Dr. Steeby's comment, practice constant vigilance at the ponds, presenting new "scarecrows," swapping devices, and inserting live humans into the picture to prevent habituation. When shooting tactics were later permitted, men with guns combined with alternating frightening strategies worked more effectively to chase the birds off the ponds. But the cormorants kept coming.

Aquaculture owners, managers, and employees on patrol reduced losses to cormorants as they scared many and shot some. But one characteristic of cormorants that plays against them on their wintering grounds is that they feed only in daylight and roost at night. We have seen that cormorants roost primarily in cypress breaks near or above water within a few miles of the farms and ponds. It's here that the birds are concentrated and vulnerable to human intrusion. Patrols invading roost sites at night or especially at dusk chase or otherwise convince the cormorants to move out of the area. Farm employees can only apply harassment tactics to disperse the birds off site. It's hoped that by visiting the roosts night after night with noisemakers and even lasers the birds will leave and settle on natural feeding areas like the banks and islands of the Mississippi River, where they "rightfully" belong.

Much of what is happening now, and has been happening in the conflict between the South's catfish farmers and wintering cormorants, may very much be moot in the years to come. The controversy revolves around channel catfish, and the fact is that where 113,000 acres of catfish farms existed in Mississippi in 2001, only 70,000 were active as of midsummer 2009, a 38 percent decrease. Alabama still claimed 21,700 active catfish acres and Arkansas 20,500. With production and sales down, only half of the state's catfish hatcheries functioned in 2009.[30]

The reasons for the decline involve practically every facet of the industry. In our present troubled economy, prices the producers got for their product dropped, catfish feed and fuel prices rocketed, the weather tanked, and catfish processors looked for larger fish, 1½- to 3-pound fish, which take three years to bring to size, as opposed to previously preferred smaller fish that reached market size in about eighteen months. The industry, in such a position, may simply not be strong enough to support the hundreds of thousands of cormorants "funneling" into the delta region each winter.

The phenomenon of Ashmole's Halo, looked at earlier, states that food supply drives population changes. It concerns availability of food at the bird's breeding grounds, but it's been shown that for cormorants the fat reserves they build in the winter better prepare and strengthen the birds for the rigors of migration, courting, and parenthood. From a practical point of view, although the ratio of cormorants to pond numbers may remain the same, fewer catfish may ultimately mean fewer cormorants for the farmers to deal with.

PART 5

The Past as a Clue to the Future

14

Fortune, Timelines, and Their Intersections

When Worlds Collide

ALWAYS, SOMETIMES, NEVER. Three simple adverbs that describe the realm of all possibilities. Also the range of choices human beings make. Heady stuff, it's true, but the phrases and philosophy do apply to what happened to the ecology of the Great Lakes and in the South's "catfish country." A simple logic works here. Opportunity *sometimes* comes to those who prepare. Conflict *always* comes to those who do not. The charter boat skippers and catfish farmers *never* looked ahead to see what was coming their way.

The idea of fortune is a phenomenon that invariably includes the words *unpredictable* and *chance* in its definition, but it also calls up the phrase "chance favors the prepared mind," attributed to Louis Pasteur, the biologist who solved the mysteries of anthrax and rabies and discovered how to make the milk we drink safe. Preparation in our case doesn't mean the farmers and captains seeing into the future but suggests not turning a blind eye to other forces in the ecosystem that might affect how they do business. Not planning for future risks and likelihoods, ignoring the possibility of change, merely makes such people and their ideas reckless, which weakens their case for being thought of as the injured parties, requiring help from state and federal agencies.

Conflicts, like those surrounding the cormorant, turned more complex when fishermen and catfish producers assumed that current circumstances would remain constant, unchanged, even as conditions changed before their eyes. Table 1 is a compilation of four separate timelines roughly tracing key events and trends for cormorants; the catfish industry; treaties, regulations, and conservation issues; and the Great Lakes fisheries. The table tells linear stories through four series of "snapshots," but nothing happens or exists on its own or is unaffected by other snapshots. These time sequences, like the conceptual "line" taught in geometry, extend both ways, reaching back into the past as well as continuing into the future. The table is merely a "segment" on that line connecting a number of points, but it is designed to illustrate some of the ecological relationships, intersections, conflicts, and collisions that affected the cormorant's standing in people's minds.

A quick glance at the table reveals a number of trends and relationships. For example, 1885 shows a steep decline in some of the major fish species in the Great Lakes, the killing of millions of wild birds, and the persecution of cormorants. Overall, there was little thought given at that time to the inherent value of wildlife other than its economic significance. The Weeks-McLean Act and the first version of the MBTA in 1913 and 1918 protected birds, but not cormorants, just as the cormorant began its modern colonization of the Great Lakes. In the 1960s, the stage was being set for cormorant conflicts as cormorant numbers crashed due to the effects of DDT, the first commercial catfish ponds were built, Rachel Carson's *Silent Spring* exposed what was happening in the environment, and Pacific salmon were introduced into the Great Lakes in hopes of rejuvenating the fishery. The year 1972 was important in that it included both the US ban on DDT and the version of the MBTA that finally protected the cormorant, permitting the huge expansion of its population. The visual presentation of these and other relationships puts them into a structured relationship.

As a clarification of how parts of a system fit together we can examine an interesting concept logicians and philosophers borrowed from the world of quantum physics: Werner Heisenberg's Principle of Uncertainty. In a simple form, outside the scope of subatomic particles, the principle says that an observer scrutinizing or surveying a system is no longer an outsider but has become an integral part of the system: the very act of observing something changes it. The idea extends to biolog-

TABLE 1

	Cormorants	Catfish Industry	Treaties Regulations, Conservation Issues	Great Lakes Fisheries
1820s				Canals allow invasive species access to lakes
1835				Sea lamprey found in Lake Ontario
1873				Alewives identified in Lake Ontario
1885	Cormorants unprotected from prosecution and feather trade		Fashion feather trade in full swing Millions of birds killed	Lake trout, sturgeon, whitefish in steep decline
1886			First Audubon Society founded in New York	
1899				Last great commercial fish harvest
1900			Lacey Act, first federal law to protect exploited wildlife	Atlantic salmon extinct in lakes
Early 1900s	Cormorant populations rise			
1913			Weeks-McLean Act ends deadly feather trade	First cormorants found nesting on Great Lakes
1918			Treaty with Canada (MBTA) protects migratory birds, but *excludes* cormorants	
1940	DDT introduced as insecticide			

TABLE I—CONTINUED

	Cormorants	Catfish Industry	Treaties Regulations, Conservation Issues	Great Lakes Fisheries
1950s				Lake trout extinct in Lake Ontario
1960s	Cormorants in steep decline	First commercial catfish ponds established	Rachel Carson's *Silent Spring,* 1962	Coho and Chinook Pacific salmon stocked
1970	Only scores of cormorant pairs remain on Great Lakes			
1971			Ramsar convention promotes wetland conservation	
1972	DDT banned in U.S.		MBTA with Mexico includes almost all birds, now *includes* cormorants	
1973			CITES enacted to regulate international wildlife trade	
Mid-1980s		Cormorants begin to impact catfish industry		
1989		28,000 cormorants overwinter in Delta Region		
1990				Angler success peaks Round goby introduced Atlantic salmon extinct in Lake Ontario

TABLE I—CONTINUED

	Cormorants	Catfish Industry	Treaties Regulations, Conservation Issues	Great Lakes Fisheries
1998	Little Galloo Island shooting		FWS issues Aquaculture Depredation Order, allows killing of cormorants at facilities	Sport fishing declines Little Galloo Island shooting
2000	Culling and egg-oiling campaigns	Asian catfish competitors enter American market in force		
2003	Cormorant numbers swell	55,000 cormorants winter in Delta Region	Public Resource, Private Resource Depredation Order allows virtually unlimited killing of cormorants	Hundreds of thousands of cormorants nest in Great Lakes area
2004			Conservation groups file suit in federal court to prohibit Public Resource Depredation Order	
2005		600 million pounds sold, 165,000 acres in production, but prices/acres decline	Court upholds Public Resource Depredation Order	
2006				600,000 breeding pairs of cormorants now on Greak Lakes
2014			Public Resource, Private Resource Depredation Orders scheduled for renewal	

ical systems as well, in that humans, like it or not, as catfish growers or fishermen, are entities in the ecological framework of the cormorant and that their works and interests conducted on such large scales alter the system. Cormorants do what they do naturally and instinctively; their behavior is what it is, fixed. They cannot plan for the future; only humans have the capacity to prepare for change. But in so many cases they don't.

A look at the commercial fishermen on the Great Lakes illustrates this idea. We saw how for years, as the yields and profits gleaned from lake trout harvests declined, fishing firms continued to send their skippers out day after day in search of lake trout, never bothering to consider what they were doing until virtually the last catchable lake trout was taken. Only then did the industry modify its gear and move on to eradicate other species. It dabbled in and tinkered with ecological issues but never acknowledged its responsibilities within the system.

Some of the changes discussed here have to do with alterations in the biodiversity of species or changes in the ratios and balances of species in the environment brought on by human, usually commercial, activities.

The repeated introduction of invasive native and nonnative species caused massive problems on the Great Lakes. The alewife was first identified in the Great Lakes, specifically Lake Ontario, in 1873. It entered the lakes probably as a result of expansion of the canal system, which created new passages into the lakes. The Erie Canal provided Hudson River alewives access to the lakes and newfound feeding and breeding territory. The alewife eventually populated the lakes to such a degree that it became the forage fish of choice for breeding cormorants at about the time the lake trout, a major predator of small fish in the lakes, became scarce. Cormorants prospered on alewife protein.

The parasitic sea lamprey also entered the lakes and staked out its claim to the Great Lakes fishery. Humans never made the connection between man, as not just an observer but an active entity in the system, and the cormorants, alewives, and lampreys. The ideas of planning and preparation never entered the picture until it was too late. The only choice remaining was to react to what had already been done.

These ecosystems were not only affected by invasive species but were also jarred when humans forcibly changed the balance and ratios of existing species. For centuries, and perhaps millennia, double-crested cormorants on their wintering grounds in the South fed on naturally occurring populations of panfish, bass, various species of forage fish, and, of course, wild populations of channel catfish. The birds roosted and fished along the flows of the Mississippi River basin and the delta. The channel cat in the wild lives, feeds, and breeds along river bottoms, undercuts, and adjacent to sunken structures, with the natural population distributed throughout the system. The river's depth, muddy conditions, and dispersal of the fish protected the channel cats from continuous heavy predation by cormorants; also schooling fish, like bluegills and gizzard shad, made more productive targets than the catfish. The ratio and balance of species in the area were disrupted when farmers in the four major catfish-producing states excavated, levied, and flooded their soy and cornfields, creating open ponds in which to stock and raise channel cats. The channel catfish was native to the Mississippi Basin, but was certainly not found in the numbers eventually cultivated in the ponds. This huge influx of biomass was not to stand untouched. Farmers constructed the first ponds in the mid-1960s, and by the mid-1970s and early 1980s resourceful double-crested cormorants were all over them, exploiting this tidy, easily fished, artificially concentrated food source. The scale of the catfish operations was eventually so large that before they realized it the producers had redesigned their own local ecosystem.

Fortune and chance are, by their very nature, "unpredictable" phenomena, but turning a blind eye and discounting the effects of human activity, the Uncertainty Principle, and Pasteur's idea of the "prepared mind" only intensifies progressive conflicts and collisions between mankind and the rest of the natural world, in our case the double-crested cormorant.

15

Concessions and Conclusions

ASIDE FROM THE various relationships we have seen occurring among cormorants, agencies, treaties, the catfish industry, and Great Lakes fisheries it's important to recognize that events and trends can repeat themselves on timelines, though not always in the same form or sequence or with the same consequences. The repetition isn't concise or regular enough to be deemed a "cycle," but does demonstrate that the past is a potential clue to the future and deserves our attention.

A significant part of the cormorant's history was its battle with persistent pesticides and other pollutants: DDT, its derivative DDE, PCBs, and the others. In reality it wasn't much of a fight. The double-crested cormorant had no resistance to the reproductive effects of these chemicals. The bird's immune system generated no antibodies to them, created no mechanisms to "wall off" the toxins, and accumulated many of them in its tissues. Concentrations of the poisons mushroomed throughout the environment and concentrated up the food chain, destroying the cormorant's eggs and reproductive potential to the point that double-crested cormorant DNA was nearly extinguished from the planet. The banning of DDT helped bring *Phalacrocorax auritus* and other fish-eating birds back from the brink, but the dispute over DDT has not ended yet.

When DDT was abandoned as a public health tool to control and eliminate malaria, in many countries it was still often applied to agricultural fields to protect valuable crops from destructive insect pests. Before long, in times as short as a one-year period, DDT-resistant strains of insects evolved, greatly reducing the effectiveness of the pesticide. Mosquito strains able to withstand DDT sprang up, survived, reproduced, and spread across broad political and geographical boundaries. As a result, DDT became less and less useful to agriculture and public health.

In many parts of Africa, the human death rate due to malaria rose quickly when DDT was banned, sometimes within the span of a year or two. Some sources claimed that thousands more people died in South Africa in 2000 until DDT was reintroduced and malaria cases were reduced by 80 percent the following year.[1]

Organizations, including the African American Environmentalist Association (AAEA) and Congress of Racial Equality (CORE), saw the abrupt banning of DDT and loss of funds for malaria control as an "eco-imperialist" act founded on poor science and racist attitudes. The AAEA's mission is to restore the use of DDT in African nations until malaria is eradicated. Quoting Paul Driessen's *Eco-Imperialism: Green Power, Black Death,* the AAEA states that American and European environmental groups forced their ecological views on African people and nations and are "preventing needy nations from using the very technologies that developed countries employed to become rich, comfortable and free of disease."[2] What the groups are claiming may have some truth in it, but they don't tell the whole story and they unfortunately ignore the scientific reality of insects acquiring resistance to pesticides.

In Laurie Garrett's 1994 *The Coming Plague: Newly Emerging Diseases in a World Out of Balance,* she relates how malariologist Paul Russell developed a plan for the eradication of malaria. In Russell's report he claims that "it takes four years of spraying and four years of surveillance to make sure of three consecutive years of no mosquito transmission in an area. After that, normal health department activities can be depended upon to deal with occasional introduced cases." Russell insisted that the time frame of eight to ten years, with no more than six years of spraying, was of the greatest importance, since longer spraying periods produced numerous strains of DDT-resistant mosquitoes. The US Congress accepted Russell's recommendation and backed an antimosquito/antimalaria program with more than twenty-three million dollars annually for the six years from 1958 to 1963. These funds made up more than 95 percent of the World Health Organization's malaria budget and large portions of the budgets of the Pan American Health

Organization and the United Nations International Children's Emergency Fund (UNICEF). It was a financial commitment equivalent to billions of today's dollars. Mosquitoes died, and the malarial parasites they carried died with them. But by 1963 malaria persisted as a major disease in many areas, even though it had disappeared in others. The program was seen by Congress as a failure since the stipulated years of spraying did not eradicate malaria. Funding from the United States soon ended, and since the United States was the only source of excess cash at the time, so did the malaria eradication program.[3]

The plans for malaria control contained more elements and twists than originally thought. Because of other legitimate needs in the nations receiving funds, some governments diverted funds and DDT resources from public health programs to agricultural cash- and food-crop applications, causing suspensions in antimalaria spraying. The longer, prolonged exposure to DDT in farmers' fields meant many more DDT-resistant insects, including both crop destroyers *and* malaria-infected mosquitoes. And so the programs in those areas failed.

Malaria, without a doubt, is still a major killer. Donald Roberts, an emeritus professor of tropical medicine, commented in the *New York Times* that malaria is "surpassing AIDS as the biggest killer of African children under age 5."[4] What CORE and the AAEA ask for now is a very limited use of DDT; sprayed on the walls of huts and homes in malaria-infested areas, rather than the past method of broadcast spraying across wide areas. Used as a "spatial repellent" DDT keeps mosquitoes from entering the hut. Mosquitoes that do enter and land on sprayed surfaces experience DDT's property as a "contact irritant," which compels them to leave the building. Even DDT-resistant mosquito strains are susceptible to DDT's repellent and irritant characteristics, and nonresistant strains do not become resistant in the process. Roberts concludes by writing, "DDT remains the cheapest and most effective long-term malaria fighter we have." The AAEA and CORE are correct in their goal of saving lives, and DDT may be the only way to achieve that goal in the short run, but at the same time they cannot ignore the environmental consequences of persistent insecticide use.

The dangers not mentioned in any of these articles and position papers are issues of poorly trained personnel, overapplication, and diverted use of DDT sprays that would introduce the persistent pesticide into the local, and in turn the global, environment, again causing long-term damage to wildlife and food chains. Also, little or nothing is mentioned in these reports of the effects on humans, young and old, of breathing and coming into contact with DDT on a day-in, day-out, walking, sitting, cooking, and sleeping basis.

Surely, concerns about DDT should involve more than the thinning of cormorant and bald eagle eggshells. As a chemical in our food and our soil and our water, DDT, in the words of Rachel Carson, has the ability to "enter into the most vital processes of the body and change them in sinister and deadly ways."[5] We dragged ourselves out of that biological quagmire once before after great cost to our wildlife and ecosystems and must be cautious of any actions indifferent to them.

Another repeating theme on the timelines with potential to affect the double-crested cormorant is the rise and fall of channel catfish propagation. The story of the catfish industry does take on more of the characteristics of a cycle as we think of one, but the scale of the repetition runs for decades and generations. The first commercial catfish ponds appeared in the mid-1960s, and we saw how the channel catfish was chosen as an alternative to row crops when prices for soybeans and corn and other crops dropped to a point where farmers no longer saw a profit. The cycle came full circle again in recent years as the prices of soy and corn, the two major components of catfish feed, soared. The combination of a weak economy and increasing production costs, coupled with the competition of imported catfish from Asia, drove many smaller catfish farmers to bulldoze at least some of their ponds and revert to their old standby, and now more profitable, crops, soy and corn. Not all the farms or ponds were covered and sown with row-crop seed, but almost 40 percent of the catfish production acreage was diverted to other uses in the first decade of this century. Today, Mississippi's farmland dedicated to these two major crops measures over a million acres each. The channel catfish may again play a major role in the South's economy and in natural history of the cormorant, but at present its importance is dwindling.

Similar conditions exist in Alabama, the nation's second-largest catfish-producing state. In 2004, Alabama claimed about 27,500 acres of productive catfish ponds, but by 2009 it had lost 5,400 of those acres, leaving 22,100 for catfish cultivation. Apparently some growers, seeing early profits and market interest in catfish products, got into the catfish game on an impulse rather than as a long-term venture. When companies such as Wal-Mart began courting Chinese catfish sources and purchasing frozen imported products right at the docks rather than from American producers, a number of smaller American growers folded their operations. Some of the ponds were merely drained and left to an overgrowth of grass and weeds grazed by the owners' cows. Others were restocked with bass and bream (pronounced "brim," also known as bluegills) sold to stock bass lakes and ponds. Cormorants raided these ponds as well, but since the margin on bass was higher, chances of turning a profit were greater.[6]

Another "cycle" worth reviewing is the fishing industry–cormorant conflict in the North. The cormorant's life on the Great Lakes, particularly on Little Galloo Island, has cycled back to what it was a few decades ago. The bloody "protest" of the charter boat skippers, which left so many dead cormorants behind and "made a local issue into a national issue,"[7] is largely forgotten by the press, but still lives in the memories of some conservation groups and, of course, on the Internet. Through the new flexibility of government regulatory agencies and programs conducted by wildlife managers and biologists, Henderson Harbor has again revisited its status as a productive, tranquil, law-abiding fishing village. Its fishermen and business owners, like Captain Mitch Franz and others, have since shed their 1998 reputation of being outside the law and bird killers, as many people across the country saw them at the time. In fact, for the past several years Franz has been chairman of the Fish Advisory Board, part of the Planning and Development Committee of Jefferson County, which includes Henderson Harbor.

By mutual agreement, consensus, and a few nods, the New York State Department of Environmental Conservation, the USDA's WS, and the charter captains have found a satisfactory population range for cormorants on Little Galloo. The target figure is now fifteen hundred to two thousand nesting pairs. Wildlife biologist James Farquhar of the NYSDEC, the overseer of the island's cormorants, said that since 1999 his goal on the island has been to oil every cormorant egg or "destroy every accessible nest" on the island. For many years the only nests left to reproduce were those constructed out of reach on the leaf-stripped limbs of dead trees. Over time, "the numbers are coming down," he said, and the program's goal is "more maintenance than reduction." His relationship with the captains is cordial; Mitch Franz says he communicates with Farquhar, and all seem content with today's status quo. According to Farquhar, the charter skippers are satisfied with cormorant numbers now but wish the reduced population had come about sooner, in fewer than ten years. Farquhar called oiling and nest destruction a "measured approach to cormorant control," suggesting that, as a compromise, the process pacified the fishermen and didn't raise the ire of conservation groups as much as a broad shooting campaign would have done. In general, the NYSDEC and the "feds" garnered "some small bits of praise" from the captains because recognition of the cormorant controversy was addressed and the initiation of some control measures created a little goodwill.[8]

To continue the good relations and in order to preserve the agreed-upon number of cormorants, the program on Little Galloo now allows 350 to 400 nests to reproduce. About 300 ground nests and 100 or so tree nests are untouched, while all other nests are oiled or destroyed. Generally the oiling and nest breakups are enough to hold cormorant numbers within the brackets of the goal, but influxes of new birds and persistent nesters that won't move on create a population exceeding the 2,000-pair mark. This is when culling again comes into play. On Little Galloo Island, yearly cormorant culls are a reality and fact of life. Annually, 400 to 700 birds are killed on the island as a "supplement to oiling." The dead birds are then either composted there or removed for burial at another site. Some of the bird carcasses might be dissected from time to time for stomach content analysis and other studies.[9]

Notice also that the higher number in the range of culled birds, 700, approaches the number of birds reportedly killed that 1998 summer night, 850 to 1,000, in the "massacre" that resulted in the federal prosecution of the Henderson Harbor captains. What makes one action a "criminal slaughter" and the other a legal cull? That decision is determined by who does the shooting and under what authority. Mitch Franz and the other men killed the cormorants out of anger and frustration at what they saw as apathy by federal and state agencies. Today the culls are done by federally funded "technicians," operating in units under federal permits, using "suppressed weapons." Suppressors are what most people think of as the "silencers" we read about and see used by secret agents and assassins. Suppressed weapons reduce the noise and flash of a firearm, which allows a continuous take of cormorants without scaring off the remaining birds. People other than technicians who shoot protected birds without permits, with these or other weapons, are then, by definition, labeled criminals. To some it seems a matter of semantics, but that is often what makes a crime a crime.

Legal or not, moral or not, the question has to be asked here about Little Galloo cormorant controls: After more than ten years of oiling and culling, has the fishing gotten better? To that there is no precise answer. Jim Farquhar said that without the existence of recent creel counts "fishing has gotten anecdotally better." In general, captains now report better catches with "more legal-size [smallmouth] bass being caught." But much depends on the success of repeated strong year-classes (fry and fingerlings) to rebuild the smallmouth bass stock back for the sportfishermen. The new period of cooperation between fishermen and wildlife managers has also fostered an agreement about the quality of fishing in the waters surrounding Little Galloo. Mitch Franz admits that the fishing has improved, but some seasons, 2009 and others, were tough years due to cooler water temperatures and generally cold, rainy weather. He does acknowledge, though, that "the yellow perch have come on really strong."[10] Yellow perch fillets, breaded and fried, are the everyday, get-a-bucketful, table fare of the Lake Ontario fisherman, so that's a statement not to be taken lightly.

On the opposite side of the cormorant issue, many conservation groups face what they seemingly consider more significant and alarming questions than cormorant control issues. Under the "where are they today" category we can review the current positions of some of the conservation groups that filed suit in federal court challenging the US-FWS's Public Resource Depredation Order, which allowed the unlimited killing of double-crested cormorants, and learn where they stand today, although after five or six years they probably will have purged their files by now.

The website of the Animal Rights Foundation of Florida contains no comments on cormorant management policies but is presently concerned with the dangerous python trade in Florida, as well as antifur and antifishing campaigns, and is in a struggle to remove circuses and their captive animals from school trips and activities. Another group, Defenders of Wildlife, again judging by its website, doesn't seem to be actively involved in cormorant issues but appears to be vigorously fighting and protesting the corporate funding and sponsorship by some sporting goods retailers (Cabela's, for one) of competitive predator hunting derbies in which hundreds of wolves are killed, an activity legally permitted in several states since wolves were delisted from protection under the federal Endangered Species Act.

The Humane Society of the United States today still reacts to the alleged impact cormorants have on the Great Lakes fisheries. It continues to claim that there is no real evidence that cormorants were the cause of the decline in the fisheries and that cormorant control is not the answer because cormorants are not the cause. According to Stephanie Boyles, the wildlife scientist for the HSUS who works with municipalities and corporations to develop humane management policies, the society has advocated against the depredation orders since 2002 because they are not attacking the problem. She said it is unethical to mischaracterize cormorants and doing so punishes the birds for what they do naturally and have done for thousands of years.[11] The society's efforts on behalf of cormorants are currently done primarily through its membership in Cormorant Defenders International, which we will discuss below.

Any discussion of bird conservation and activism would not be complete without including the National Audubon Society. We saw earlier how some of the original Audubon societies worked to end the brutal fashion feather trade and to educate the public in the ways of wildlife. Audubon today is mostly concerned with maintaining visible populations of wildlife, preventing extinctions, reversing population declines, and keeping common birds common. The management of species currently increasing their numbers, like our cormorant, is not among Audubon's highest priorities. This prioritizing, however, does not exclude cormorants from Audubon's outlook. The author spoke to Dr. Greg Butcher, Audubon's director of bird conservation, searching for a "policy statement," but gained other insights along the way. Dr. Butcher asserts that shooting wild birds without concrete, proven reasons is taking a "cavalier attitude" toward wildlife and allows for the overzealous use of depredation orders. In the South, the catfish ponds are essentially an "attractive nuisance," supported by the operational side of USDA/WS, which Butcher says is biased more toward the financial interests of agricultural producers than wildlife. He also suggests that a better alternative to expanding aquaculture facilities that are not run in very environmentally sound ways would be to restore natural fisheries and harvest them at sustainable levels.[12] Dr. Butcher is correct that new natural fisheries might take the pressure off double-crested cormorants and the ponds, but new regulations and quotas would be required to avoid the "tragedy of the commons" and the eventual persecution of competing cormorants feeding on the same free-swimming fish stocks.

Aside from conservation groups working their way through the tangle of cormorant concerns, several states also continue to face cormorant issues. New York manages Little Galloo and other colony islands, and other states bordering Great Lakes waters, Wisconsin, Michigan, and Minnesota, are looking at ways to control cormorant populations. And among them, at least one state also conducts cormorant control programs on other lakes separate from "Fifth Coast" shores.

In many parts of the Midwest, the yellow perch, a typically plentiful panfish, sought more by the area's recreational fishermen for its skillet value than its fighting ability, is the focus of many cormorant-human conflicts. In Wisconsin, cormorant numbers increased by almost 25 percent annually between 1973 and 1997 to reach a total of more than 10,000 nests. More recent tallies show about 18,000 nests and 42,000 birds, including nonbreeding adults. Most of the birds are concentrated on Lake Michigan, near Green Bay. Area fishermen, in their minds, link the decline in the yellow perch fishery to the increasing double-crested cormorant population. Reports show that the birds do feed on yellow perch, but not at all exclusively, as they prey on several other species as well, so cormorants may not be the ultimate culprits. Oiling of cormorant eggs began in 2005, resulting in a gradual decrease in the bird's population. A Wisconsin Department of Natural Resources wildlife biologist wanted to reduce the area cormorant population to less than 6,000 nesting pairs, but since "cormorants are a reflection of how much fish biomass is out there" another incursion by still another invasive fish species could provide an impetus for an upswing in cormorant numbers.[13] And as we've seen, the idea of more cormorants is not necessarily a bad thing, in that through the species' sensitivity to environmental chemicals it is a living monitor of the ecosystem's health.

Still another potential hot spot for cormorants is the Les Cheneaux Islands region of Lake Huron in Michigan. Again it involves the collapse of the yellow perch fishery. In 2003 the *Delta Farm Press,* a magazine concerned with agriculture issues in the South's Mississippi Delta region, ran an online article on what it called the "common misery" shared by Michigan and the delta: an overabundance of cormorants. In it, the point was made that "if the Delta's cormorant problem is bad, Michigan's problem is horrific." The same article argued that the more than eleven thousand double-crested cormorants on Michigan's islands had destroyed the area's perch fishery and that breeding adults in the North were sending multiples of their number back into the South to feed on farmed catfish every winter. The magazine essentially blamed Michigan's lack of action for the cormorant predation of their ponds.[14] Michigan's cormorant programs may be a factor in the delta's winter cormorant population, but the catfish industry's denial of the delta's open invitation of unprotected catfish ponds as an initial instigator cannot be ignored either. If the ponds weren't there . . .

A 2008 press release issued by the International Association for Great Lakes Research cited a study by the Michigan Department of Natural Resources, which maintained that cormorant predation was the most influential factor affecting the yellow perch fishery.[15] The study was, however, only an examination of "trends" in perch reproduction, weather variables, water levels, fishing activity, and cormorant predation. The conclusion drawn in the study has at least one serious flaw: it was made via an elimination process, regression analysis, rather than through positive, direct evidence. The study was a survey of possible relationships rather than a set of verifiable experiments and observations. It was decided that the cause of the perch decline was not in its reproductive activity, not in the weather and water level trends, and not in fishing and angler activities. That left only one reason for the yellow perch decline: the cormorants. The study makes the large first assumption that it examined *every* factor affecting the fishery and eliminated each one, except predation by cormorants. What about pollution, development, erosion, eutrophication, long-term effects of overfishing, and other influences not mentioned? In discussing reproduction and fishing activity, a suggestion made to create a closed season in the bays to protect spawning yellow perch was rejected by state wildlife biologists because they didn't want to put pressure on anglers. The study and press release also failed to mention if any diet studies were conducted on the area's cormorants. What was in their bellies? Those findings would provide the direct, positive proof of the effects of heavy cormorant predation on yellow perch numbers, which trends, theoretical relationships, and elimination processes do not.

Minnesota is another state that today continues its attempt to manage and control cormorants. Minnesota borders Lake Superior on its northwest shore, but its publicized cormorant predicament lies inland from there, at Leech Lake, a legendary Midwest fishing hole. The cormorant-human conflict proceeded along the same lines as on the Great Lakes, but in this case the declining Leech Lake fishery not only included the yellow perch but also the walleye, a larger, toothy, tasty relative of the perch with enormous popularity among northern anglers.

The cormorant nesting site in question is Little Pelican Island, owned and managed by the Leech Lake Band of Ojibwe, a Native American tribe. The decline in walleye began in the early 2000s, and cormorants were the suspected walleye bandits. In 2007, both lethal and nonlethal measures were applied on the island with about 2,600 birds shot and another 348 removed from sites through egg oiling, with all work done by WS personnel. That number brought the four-year culling total to more than 11,000 birds. As the population of double-crested cormorants was drawn down, the walleye fishery recovered, which was taken as a true sign of success and validation of the culling programs. Though not as publicized as the seeming culling success, what also aided the recovery was an intensive walleye stocking program and tighter restrictions on anglers taking fish home. So, likened to the "conclusive study" based on theoretical relationships, the state and tribal biologists on Leech Lake failed to isolate the variables, such as an unexplained natural walleye reproductive decline and overfishing (remedied by stocking and angler restrictions), instead outright excluded them, and simply found the cormorants to blame.

Many of the stakeholders on the lake, fishermen and resort owners, were happy with the artificial ratio between birds and fish created by the shooting culls and oiling, but there were still some who would not be content until every cormorant was dead and gone. There were diet studies done on Leech Lake cormorants, which when completed showed that only 1 to 3 percent of the bird's food intake consisted of young walleye. Linda Wires, a biologist we encountered earlier and will again, who has studied Minnesota cormorants extensively, was quoted by the *Minnesota Post* as saying, "In fishing communities, there is just such a low tolerance, almost zero tolerance, for cormorants, it doesn't seem to matter much what the data says." She added that the agencies will continue to "manage" cormorants because of the public exerting such great pressure on them.[16] The organization Audubon Minnesota takes a similar position, stating that setting bird management goals based on fishery objectives is "contrary to good bird management" and that to "keep common birds common" is at odds with the philosophy of both the USFWS and the Minnesota Department of Natural Resources.[17] It seems the persistent thread in which perception supersedes reality continues to run through the entire story of the cormorant, including state and federal agency versions of it.

Earlier we came across Cormorant Defenders International, a group very active today in cormorant conservation issues. Located in Toronto and working primarily in Canada, CDI makes the claim that "a great deal of misinformation about cormorants has been spread by anglers and wildlife managers fueling an organized war against the birds on both sides of the US/Canada border."[18] The organization was founded to respond to the assault on cormorants in the Great Lakes area and to react to the "erroneous claims made about them." It works to convince politicians, anglers, and wildlife managers that the return of cormorants and their recolonizing efforts are natural processes. It counters the objection that cormorants destroy vegetation by saying that changes in plant groups occur in bird colonies around the globe and that plant progression is also a natural process signifying evolving ecosystems.

Cormorant Defenders International documents and records culling and egg-oiling programs in Canada to make the public aware of what actually happens during these actions. Julie Woodyer, campaigns director for CDI, who also works to protect captive animals, says that the culls are "horrifically cruel," with "one in ten birds left to bleed out" or die of their wounds up to a week later.[19] The organization's literature maintains that culls are expensive and only produce temporary results unless they are performed on a continuous basis. Killing the birds has no effect on an area's appeal to cormorants, so they will return to these nesting sites year after year regardless of shooting programs. Cormorant Defenders also asserts that massive culling actions may drive off other, more timid bird species, such as the great blue heron, which nests alongside cormorants and has a less powerful nesting drive.[20]

Woodyer also pointed out that some university students are presently studying the cruelty issue in culling and the idea of "social intolerance" toward cormorants regardless of the protection afforded them in parts of Canada. The country does have laws to protect migratory birds similar to those of the United States, but when the MBTA was initially signed with Canada in 1972, cormorants were not yet included. Therefore, the cormorant has no *federal* protection in Canada, leaving that option up to the various provinces and creating regional differences in protection and management measures.

A typical area of concern in Canada is Middle Island, a tiny island in Lake Erie and a world-renowned bird sanctuary. Cormorant Defenders International fought in Canadian federal court to end the culling of cormorants on the island begun in 2008. According to Woodyer, most of Middle Island was owned by the Nature Conservancy until it was turned over to the Canadian federal government. She argues that the government's first action after it acquired the Middle Island property was a plan to eliminate the cormorant from it.[21] As detailed on its website, as a result of CDI's challenge, fewer than 250 birds were killed instead of the many thousands originally planned. About 1,600 were killed in 2009, and the goals for 2010 were uncertain.[22]

Another locale of concern to CDI is Tommy Thompson Park, a man-made peninsula close to downtown Toronto extending into Lake Ontario. Composed of the diggings removed to create the subway system, it is now home to a number of varied plant communities and large numbers of nesting double-crested cormorants. Its proximity to the city created a very political issue and, on a practical basis, eliminated any shooting programs to control the bird's population. It just wasn't good publicity or politically correct to shoot wild birds in full view of the public. As a compromise, cormorants were allowed to nest unmolested on one portion of the peninsula, while lively human activity would keep the birds from colonizing the remainder. As a result, the park has become a tourist attraction for professional and amateur botanists, birdwatchers, and other naturalists. The fishing close to the peninsula is reported to be as good as or better than it was before double-crested cormorants adopted the point as a rookery.[23] And according to CDI's director, Ainslie Willock, a webcam is scheduled to be activated at the nesting site so that Internet users around the world can watch Toronto's cormorants.[24] Conservationists and wildlife managers might look at Toronto's Tommy Thompson Park and note it as a model for future successful cormorant management—without firing a shot.

Because double-crested cormorants are migratory birds they pay no attention to international borders. About 80 percent of cormorant nesting takes place on the Canadian side of the border, so unsurprisingly, many parties and stakeholders are interested in what happens in Canada. Several isolated comments suggest that US agencies have tried to influence cormorant control policies in Canada, but Diane Pence, chief of the USFWS's Division of Migratory Birds in the Northeast, said that the USFWS is charged with overseeing programs, not initiating them, and that it had had no contact with Canadian agencies. Pence noted also that WS, under the USDA, is mandated to solve wildlife damage problems, not formulate plans. She suggested that any contacts made with Canadian agencies flowed from individual states and private groups.[25] One such group, the St. Lawrence County Cormorant Control Working Group, an informal group of concerned sportsmen and others, was convened by the county legislative chairman. According to James Farquhar, NYDEC wildlife biologist and adviser to the working group, representatives of the group made inquiries to Canadian officials regarding their cormorant control programs since the St. Lawrence River/Seaway system offers nesting sites for a great many cormorants that feed on both sides of the waterway. The return communications were informal and limited, producing little, if any, cooperative planning.[26] The lack of success in communication with Canadian officials should not be a surprise. Much of what is said and done about cormorants on either side of the border is subject to budgetary concerns and priorities, which are sorely limited in many of today's economies.

On the continental scale, beyond the territories of North America, the double-crested cormorant of the Americas is not the only cormorant species to be maligned and hated in the world. A related cormorant, within the same genus as the double-crested, *Phalacrocorax carbo,* the great cormorant, also known as the black shag, has found itself in trouble in many parts of Europe. One writer in the *Irish Times* described the great cormorant as the bird of November, "that gothic black silhouette, the wings like a broken umbrella hung out to dry."[27] Ireland, France, Germany, Norway, and other countries have instituted cormorant study and control programs. Across Europe, groups like Intercafe, the Interdisciplinary Initiative to Reduce Pan-European Cormorant-Fisheries Conflicts, came into being to regulate cormorant population growth. Problematic cormorants in Europe are characterized along with grey seals as "over-conserved predators needing 'management.'"[28] The term *over-conserved* reads and sounds much like the description *overabundant* used in the United States by anticormorant activists. The titles of articles and studies in Europe also mirror those in the United States: "The Cormorant: The 'Black Plague' or an Example of Successful Species Conservation?"[29] Similar birds, similar perceptions?

Europe faced the same types of inconsistencies concerning cormorants as did the United States. In Ireland, until 1976, with the passage of its Wildlife Act, the Republic paid bounties for the killing of cormorants. When the bounty was finally lifted cormorant numbers quickly rebounded. And again, as in the United States, in addition to the end of their persecution (the MBTA), cormorant growth was tied to massive sport fish stockings of angling lakes, which created a sizable new food source for the hungry birds. Also parallel to what happened on the Great Lakes, predatory fish, namely, lake trout, Atlantic salmon, and whitefish in the United States and predatory pike in the Irish Republic, were taken out of the equation and ecosystem, permitting forage fish to "bloom" and provide a varied menu for already expanding cormorant populations. As a result, great cormorants today number about half a million individuals across Western Europe.[30]

Parallels in cormorant conflicts abound as deep into the US-European comparison as one cares to delve. Regional, national, and international attitudes were always important factors in developing cormorant policies. And since these cormorants, like our double-cresteds, do not read maps and have no respect for political boundaries along their migratory paths, any plan that is to be at all effective must be applied across international borders. The European Union is composed of twenty-five member nations, not all with the same attitudes toward cormorants, making a required consensus difficult to achieve; if just one member dissents, action is deferred. The Netherlands opposes the culling of cormorants in any number or form. France, on the other hand, expresses no hesitation in shooting forty thousand cormorants every year.[31] Such differences likely assure that a continentwide policy on cormorants is not in Europe's immediate future.

On the European continent the ideas of who is responsible and who sets population goals reflect the same questions faced in the United States. Conservation groups like BirdWatch Ireland want their say as to whether culling will actually protect the fisheries, and fisheries managers and other agency officials have their own agendas and quotas in mind. A 2008 press release from the Helmholtz Centre for Environmental Research (UFZ) holds the United States up as a model of successful cormorant management. The UFZ boldly asserts that the "central authority," the USFWS of the "Ministry" of the Interior, produced a two-hundred-page cormorant management plan, "which is now being applied religiously."[32] Though meant as praise for America's doggedness, the compliment does oversimplify and exaggerate the success of the multilayered cormorant management programs we have examined thus far. Only on paper does it work so well.

On the other side of the planet, in China, the world the cormorant lives in also faces many changes. Traditions in China, especially those associated with harvesting natural resources, have existed for centuries and perhaps millennia. Cormorant fishing is one of them. Earlier we saw how Chinese fishing masters raised and maneuvered cormorants to catch fish for their families and for sales in the market. Cormorant fishing was both productive and profitable, but in modern China many families have abandoned the tradition for a number of reasons, some of which also exist in the United States and across Europe. Unique to China in name, but not in its implication, the Cultural Revolution lessened the sanctity, significance, and truth of drawing from nature by fishermen and farmers as practiced in Taoism and Confucianism, replacing them with the concepts of oppressive classes and struggles with nature. (We will examine the loss of nature later.) Other factors common to China, the United States, and Europe involve the regular pollution of waterways with industrial and agricultural wastes, diseases and weakened eggshells among cormorants, runoff from logging activities, and, as industrialization opened the Chinese economy to the outside, the introduction of nonnative, exotic fish species.[33] Like the alewife and round goby of the Great Lakes, these invading species outcompeted native species, making cormorant fishing less and less practical and its future as a tradition very uncertain.

Also uncertain may be the work of conservation groups in protecting cormorants. A search of the websites of thirteen other animal rights/wildlife protection groups turned up few if any responses to inquiries concerning cormorants or cormorant controls. Many groups evidently consider current cormorant questions not important enough to gain "active involvement" status. In light of other wildlife alarms, alerts, and crises, cormorant conflicts and control measures have lost their standing and priority and seem to have fallen below the conservationists' radar. A general agreement seems to be that there are just too many cormorants in circulation to consider them threatened, endangered, or of importance as a true conservation issue.

Linda Wires, a research fellow at the Department of Fish, Wildlife and Conservation Biology at the University of Minnesota, who we encountered earlier, is among those who look at cormorant populations and controls as valid conservation issues. Wires admits that cormorants are a "really hated species" and that "tolerance for cormorants is very, very low."[34] A paper cowritten by Linda Wires and Francesca Cuthbert suggests that the birds be granted cormorant "safe zones" where "human interests are not allowed to influence cormorant numbers."[35] Wires added in a phone conversation that the creation of "safe zones" is not very likely, since even in US designated wilderness and wildlife areas set aside for the specific protection of wildlife, large numbers of birds, including cormorants, are being controlled, meaning shot, by government technicians.[36]

Wires also mentioned the oft-heard remark that conservation groups and wildlife control agencies are working toward the same goal, but "conservation is not the same as control." Today, conservation organizations do not often see it that way, primarily because cormorants are at "somewhat natural populations," but they are important because they "reflect heavily on the condition of the environment," much like African elephant numbers are indicators of how well that ecosystem is doing.[37]

Another observation that can be made about cormorant conservation and control issues concerns the debate between aquaculture interests and conservation groups. The two parties continually antagonize each other, fighting the battle from fixed positions with the same language and the same arguments, virtually deaf to the other's repeated reasoning. Conservation group spokespeople sometimes deal in generality and simplification when discussing aquaculture issues. Catfish producers complain of hundreds if not thousands of cormorants raiding one pond after another, their slim prohibitive profit margins, foreign competition, and increasing numbers of cormorants migrating south each winter. In an almost ritualized and rehearsed series of responses conservation groups describe various nonlethal measures as if the producers had never heard of or employed them and tend to ignore the reality that all those measures work for only short periods of time before the birds habituate to them. The groups also disregard the fact that cormorants adjust even to the effects of periodically rotating and alternating the different deterrents. It seems that conservation groups rarely address the reality that producers work on shoestring budgets with small labor forces paid for through those slim profit margins on their products. A counterpoint was offered by the HSUS in the voice of Stephanie Boyles. She criticized the catfish farmers for opening their ponds to cormorants with a virtual invitation to feed there without protecting them, as they do, for instance, in Israel and Hong Kong, with mechanical devices that limit the birds' access to the ponds. Boyles also noted that such protections show that "we can use our brain to reduce losses to an acceptable level."[38]

But, catfish producers cannot be held blameless either. The producers and some southern wildlife biologists antagonize the conservationists by referring to them collectively as "those animal rights people," hinting of an extremist point of view. The producers also damage their own position in that they have come to depend so heavily on the efforts of federal wildlife biologists and WS personnel without having turned portions of their earlier profits back into their businesses to better protect their ponds, essentially unguarded "bird feeders." The producers and their spokespeople also reestablish the "line in the sand" scenario every time they ask in the media "What good are they anyway?" meaning "What right do cormorants have to live?"

The final paragraphs of the closing chapter of this book are an excellent place to take a last look at what we have discovered and observed, and what others have written about our topic.

As to the Great Lakes cormorant controversy, things are quieter now, and have settled into a convoluted balance, but departures from the standard are never far off. Climate change, though a political football, looks over the modern fisherman's shoulder. The economy runs through predictable and unpredictable cycles. Fish populations swing through extremes. And not all reversals in fish numbers are products of nature. Rarely have fishermen as a group, commercial or sport, ever admitted that they take too many fish. An older fisherman the author met once at a local lake proclaimed that he couldn't understand what had happened to the yellow perch in the lake. Years ago he would fish through the winter ice in the morning and catch "a coupla hundred," which he would take home, clean, and freeze. Later in the day he would hit the lake again and haul home "a coupla hundred more." And he didn't know where the yellow perch went? Margaret Beattie Bogue writes in her book *Fishing the Great Lakes; An Environmental History, 1783–1933,* that "the fishing industry of both oceans and lakes exerted a dominant influence on policy makers and justified its actions by using the oft-repeated argument that the waters are so vast and the fish so fecund that depletion is impossible." We now know the truth of that statement. She concludes with a statement, again about the fish but obviously related to the success of our cormorants: "One of the major challenges facing all the peoples of the world today remains as it was a century ago: to save the fish."

In speaking to Stephanie Boyles of the Humane Society, she mentioned the philosophy of Aldo Leopold, a conservationist known for his concept of the "land ethic" and his book *A Sand County Almanac,* a collection of forty-one conservation essays. Leopold's book has been compared to Rachel Carson's *Silent Spring* and Henry David Thoreau's *Walden* in its influence on conservationist thinking. The term *land ethic* refers to Leopold's vision of a community that in addition to people also includes the nonhuman components of the soils, waters, plants, and animals. His concept is that this broad community should be dealt with in an ethical way, upholding the highest of human values, and since our economic health is tied to the health of the environment, humans should have some sort of personal connection to the land. It is this personal relationship that helps us understand actual relationships in nature, including the niche carved out by the double-crested cormorant. Aldo Leopold's land ethic, like all ethics, helps us to evaluate choices, in this case relating to the plants, animals, and waters of the community. It affirms "their right to continued existence, and, at least in spots, their continued existence in a natural state."[39]

One of the other threads running through this book is the idea that perception often supersedes and supplants reality when humans deal with cormorants. Shauna Hanisch, a former USFWS wildlife biologist and doctoral candidate studying human attitudes in wildlife management at Michigan State University, admitted in an interview that the cormorant does have definite "charisma issues" and suffers because of its high visibility. When asked about the conflicting missions of agencies to both protect and control wildlife, she commented that "the contradiction is not all that great" because it's a matter of balancing nature and conflicts. She explained that if damage issues are not addressed then the public's image of wildlife is damaged. On the protection side of the issue, Hanisch added that if there is no actual damage then there is no reason to kill wildlife.[40]

Many of the perceptions and misconceptions we have about wildlife are often driven by attitudes planted in our minds at earlier times in our lives. Richard Louv, in his *Last Child in the Woods: Saving Our Children from Nature-Deficit Disorder,* discusses how an appreciation of nature can change the way humans look at the world. Without such an appreciation, the protection of a unique animal such as the double-crested cormorant cannot be initiated or sustained. Louv explains, "[W]hen I use the word 'nature' in a general way I mean natural wilderness: biodiversity, abundance. . . . Most of all, nature is reflected in our capacity for wonder."[41]

Louv postulates a "Third Frontier," which follows the first frontier of explorers Lewis and Clark and the second frontier of conservationist Teddy Roosevelt. The Third Frontier is characterized by a loss of interest in and ties to farming, as well as a disinterest in and misconception of earlier heroes associated with the outdoors such as Daniel Boone and Davy Crockett: "A generation that came of age wearing buckskin jackets and granny-dresses is now raising a generation for whom all fashion—piercing, tattoos, and all the rest—is urban."[42] Much of what young people learn about nature comes from movies about marching penguins and documentaries closing with gloomy conservation cautions. In earlier times, before specialized cable networks and urban trends, and probably before the popularity of television itself, the outdoors was not a passive place to observe from a distance; many writers, teachers, scout leaders, and parents held the "unquestioned belief that being in nature was about *doing* something, about direct experience—and about not being a spectator."[43] Firsthand outdoor experiences, even at the simplest level, counted for something.

Without intentionally playing "that younger generation" card, it is easy to see everywhere our detachment from nature. Fewer and fewer people can discuss where or how their food originates: beef from steers, ribs from hogs, tuna from the sea, mayonnaise from eggs and oil, pickles from cucumbers, "nuggets" from real chickens, and so on. Food comes from supermarkets, it's true. They make life easier, but at the same time they contribute to the gap between nature and humans, between perception and reality.

This loss of interest in nature may fall into the cause-effect poser, but both sides of the question offer a more and more acceptable excuse for a simple lack of curiosity. What happens when there is no longer a sense of amazement? Curiosity and amazement are the driving forces for understanding Louv's concept of nature: "biodiversity, abundance." As to cormorants, they are a part of nature, part of "the wilderness," however one wishes to define it, and deserve *not* to be punished for what they do or for their own success. Without these two emotional components of our psyche, curiosity and amazement, the very idea of systematically shooting tens of thousands of wild cormorants to maintain an urban provincial park, or even the killing of cormorants on Little Galloo Island just to make a political point, becomes tolerable. Conquering "nature-deficit disorder" in humans of all ages is essential to understanding and appreciating the true balance of nature. And in our case it means aligning the conflicts among the various combinations of stakeholders in our minds: cormorants versus fishermen, cormorants versus catfish farmers, cormorants versus development, fishermen versus government, conservationists versus fishermen, conservationists versus government, catfish farmers versus government, and all the other combinations and permutations.

In the minds of many people the double-crested cormorant is a sinister bandit and thief, stealing fish under the very eyes of humans who claim to be the sole owners of the bounty. One crime *Phalacrocorax auritus* did commit through its own biological success and perseverance is that it became common, familiar and ordinary. And with familiarity came contempt, as the old bromide goes, and also came a loss of the sense that cormorants *are* amazing.

Notes

PREFACE

1. Levy, "Pennsylvania Is Latest State to Put Cormorants on a Hit List," 2006.
2. Mosedale, "Most-Hated Birds in the World," 2008.

CHAPTER 1

1. Revkin, "A Slaughter of Cormorants in Angler Country," 1998.
2. Mitchell Franz, telephone interview with author, January 29, 2010.
3. Mitchell Franz, telephone interview with author, January 29, 2010.
4. Lydecker, "Game Fish Controversy Takes Wing," 2000.
5. Diane Pence, personal communication, March 16, 2010.
6. Ronald Ditch, quoted in Revkin, "In a Fishing Hamlet, No Grief for Cormorants," 1998.
7. Fish and Wildlife Branch, "Review of the Status and Management of Double-Crested Cormorants in Ontario," 2006.
8. Watts and Bradshaw, "Population Explosion by Double-Crested Cormorants in Virginia," 1996.
9. Scott and Sweet, *The Submarine Bird*, 1980.
10. Berger, "Recycling for the Birds," 1994.
11. Scott and Sweet, *The Submarine Bird*, 1980.

CHAPTER 2

1. Friedman and Gilliland, "The Fisherman and the Cormorants of Udrost (a Norwegian Folktale)," 2009.
2. Frame, "The Cormorant," 1960. Used with permission, *Saturday Evening Post*.

3. "The Fishing Cormorant," 1886.
4. Scott and Sweet, *The Submarine Bird*, 1980.
5. Dennis, *The Living Great Lakes*, 2003.

CHAPTER 3

1. Adoutte, "The New Animal Phylogeny," 2000.
2. Chiappe, *Glorified Dinosaurs*, 2007.
3. Feduccia, *The Origin and Evolution of Birds*, 1999.
4. Chiappe, *Glorified Dinosaurs*, 2007.
5. Shipman, *Taking Wing*, 1998.
6. Chiappe, *Glorified Dinosaurs*, 2007.
7. Chiappe, *Glorified Dinosaurs*, 2007.
8. Chiappe, *Glorified Dinosaurs*, 2007.
9. Feduccia, *The Age of Birds*, 1980.
10. Wilford, *The Riddle of the Dinosaur*, 1985.
11. Chiappe, *Glorified Dinosaurs*, 2007.
12. Feduccia, *The Age of Birds*, 1980.
13. Shipman, *Taking Wing*, 1998.
14. Shipman, *Taking Wing*, 1998.
15. Ritter, "Dinosaur Find Raises Debate on Feather Evolution," 2009.
16. Wilford, "Scientists Discover a Dinosaur Clad in Flightless Feathers," 2002.
17. Henderson, *Birds in Flight*, 2008.
18. Henderson, *Birds in Flight*, 2008.
19. Feduccia, *The Origin and Evolution of Birds*, 1999.

CHAPTER 4

1. Feduccia, *The Age of Birds*, 1980.

CHAPTER 5

1. Board and Scott, "Porosity of the Avian Eggshell," 1980.
2. "Eggshell Texture and Structure."
3. "Organochlorines."
4. "Organochlorines."
5. "DDT Extoxnet (dichlorodiphenyltrichloroethane)."
6. "DDT Pesticide Molecule."
7. "History of Malaria during Wars."
8. Russell, "Chapter I Malaria/DDT."
9. "DDT: An Introduction."
10. "Chemical Demand Outstrips Output," 1947.
11. "DDT (dichlorodiphenyltrichloroethane)."
12. "Section II: Uses of DDT," 2006.

13. "DDT Ban Takes Effect," 1972.
14. Kehoe and Jacobson, "Environmental Decision Making and DDT Production at Montrose Chemical Corporation of California," 2003.
15. Carson, *Silent Spring,* 1962.
16. Carson, *Silent Spring,* 1962.
17. Carson, *Silent Spring,* 1962.
18. Schuler, "Team Finds Key Mechanism of DDT Resistance in Malarial Mosquitoes," 2008.
19. "DDT Pesticide Molecule."
20. Fahrenthold, "Male Bass across Region Found to Be Bearing Eggs," 2006; Pinkey, "Endocrine (Hormone) Disruption in Fish," 2006.
21. Fry, "Unexpected Side Effects of Chemicals Acting as Hormone Mimics," 1995.
22. "Cormorants Have Fouled Island with Pollutants, a Report Says," 1999.
23. Rustem, *Double-Crested Cormorants in Michigan,* 2005.
24. "DDT Extoxnet (dichlorodiphenyltrichloroethane)."
25. Dunn, "Drastic Deformities."
26. "Toxic Contaminants in the Environment."
27. "The Rise of the Double-Crested Cormorant on the Great Lakes," 1996.
28. Dunn, "Drastic Deformities."
29. Custer et al., "Organochlorine Contaminants and Reproductive Success of Double-Crested Cormorants from Green Bay, Wisconsin, USA," 1999.
30. Dunn, "Drastic Deformities."
31. "DDT Extoxnet (dichlorodiphenyltrichloroethane)."
32. Carson, *Silent Spring,* 1962, 25.
33. "The Story of *Silent Spring.*"
34. McLaughlin, "*Silent Spring* Revisited."
35. "DDT Ban Takes Effect," 1972.
36. Cox, "Pesticides and Birds," 1991.
37. "DDT Ban Takes Effect," 1972.
38. "DDT Extoxnet (dichlorodiphenyltrichloroethane)."
39. "New DDT Report Confirms Data Supporting 1972 Ban, Finds Situation Improving," 1975.
40. "25 Years after DDT Ban, Bald Eagles, Osprey Numbers Soar" (press release), 1997.
41. Korfanty, Miyasaki, and Harcus, "Symposium on Double-Crested Cormorants," 1997.

CHAPTER 6

1. Ashworth, *The Late, Great Lakes,* 1986.
2. Ashworth, *The Late, Great Lakes,* 1986, 4.
3. Bogue, *Fishing the Great Lakes,* 2000, 4.
4. "Great Lakes Fact Sheet."

5. Ashworth, *The Late, Great Lakes*, 1986, 94.
6. "Battle of the Shipbuilders."
7. Emerson, "History of Henderson, NY, from Our County and Its People."
8. "The War of 1812."

CHAPTER 7

1. Bogue, *Fishing the Great Lakes*, 2000.
2. Bogue, *Fishing the Great Lakes*, 2000, chap. 2.
3. Ashworth, *The Late, Great Lakes*, 1986, 241.
4. Bogue, *Fishing the Great Lakes*, 2000, 5–9.
5. Bogue, *Fishing the Great Lakes*, 2000, 49.
6. "Decline and Collapse of Commercial Fishing."
7. Ashworth, *The Late, Great Lakes*, 1986, 117.
8. Hirsh, "Guest Column: Fish Shares and Sharing Fish," 2009.
9. Bogue, *Fishing the Great Lakes*, 2000, 159.
10. "Great Lakes Fish and Fishing."
11. "Michigan Commercial Fisheries Marketing and Product Development."
12. Pistis and Lichtkoppler, "Michigan's Great Lakes Charter Fishing Industry in 2002," 2003.
13. Wires and Cuthbert, "Historic Population of the Double-Crested Cormorant," 2006.
14. Fish and Wildlife Branch, "Review of the Status and Management of Double-Cressted Cormorants in Ontario," 2006.

CHAPTER 8

1. Gaston, Ydenberg, and Smith, "Ashmole's Halo and Population Regulation in Seabirds," 2007.
2. Ashmole, "The Regulation of Numbers of Tropical Oceanic Birds," 1963.
3. Gaston, Ydenberg, and Smith, "Ashmole's Halo and Population Regulation in Seabirds," 2007.
4. "The Rise of the Double-Crested Cormorant on the Great Lakes," 1996.
5. Neuman et al., "Spatial and Temporal Variation in the Diet of Double-Crested Cormorants Breeding on the Lower Great Lakes in the Early 1990s," 1997.
6. "Thiamine Deficiency Complex and Fish Mortality," 2006.
7. Sharp, "Alien Invasion," 2007.
8. Johnson, "The Effects of Egg Oiling on Fish Consumption by Double-Crested Cormorants on Little Galloo Island, Lake Ontario, in 2003," 2003.
9. Johnson, "Double-Crested Cormorant Studies at Little Galloo Island, Lake Ontario in 2008," 2008.

CHAPTER 9

1. Ronald Ditch, quoted in Revkin, "In a Fishing Hamlet, No Grief for Cormorants," 1998.
2. Ronald Ditch, quoted in Revkin, "In a Fishing Hamlet, No Grief for Cormorants," 1998.

3. Ronald Ditch, quoted in Revkin, "In a Fishing Hamlet, No Grief for Cormorants," 1998.
4. David Miller, quoted in Revkin, "A Slaughter of Cormorants in Angler Country," 1998.
5. David McCrea, quoted in Revkin, "In a Fishing Hamlet, No Grief for Cormorants," 1998.
6. Diane Gamble, quoted in Revkin, "In a Fishing Hamlet, No Grief for Cormorants," 1998.
7. Christopher Hanley, quoted in Revkin, "In a Fishing Hamlet, No Grief for Cormorants," 1998.
8. Stephen Fort, quoted in "Grand Jury to Investigate Killing of Cormorants," 1998.
9. Revkin, "A Slaughter of Cormorants in Angler Country," 1998.
10. New York State Department of Environmental Conservation, 2005.
11. Revkin, "A Slaughter of Cormorants in Angler Country," 1998.
12. Revkin, "A Slaughter of Cormorants in Angler Country," 1998.
13. Revkin, "A Slaughter of Cormorants in Angler Country," 1998.
14. "Grand Jury to Investigate Killing of Cormorants," 1998.
15. Mitchell Franz, personal communication, January 29, 2010.
16. Revkin, "9 Men Plead Guilty to Slaughtering Cormorants to Protect Sport Fishing," 1999.
17. Mitchell Franz, personal communication, February 23, 2010.
18. Revkin, "9 Men Plead Guilty to Slaughtering Cormorants to Protect Sport Fishing," 1999.
19. Thomas, "Conservation Management Takes a Wild Turn."
20. Thomas, "Conservation Management Takes a Wild Turn."
21. "Cormorants Culprits of Contamination," 1999.
22. Walkom, "Shooting Cormorants over Dead Trees Raises Suspicions about Liberal Motives," 2004.
23. Rebhahn, "Cormorant Controls Encouraged," 2001.
24. Lydecker, "Game Fish Controversy Takes Wing," 2000.
25. Sharp, "Alien Invasion," 2007.
26. Walkom, "Shooting Cormorants over Dead Trees Raises Suspicions about Liberal Motives," 2004.
27. Scrivener, "30,000 Cormorants Destroying Lakeside Park," 2009.
28. Walkom, "Shooting Cormorants over Dead Trees Raises Suspicions about Liberal Motives," 2004.
29. Scrivener, "30,000 Cormorants Destroying Lakeside Park," 2009.
30. Oosthoek, "Cormorant Debate," 2009.
31. Wires and Cuthbert, "Historic Population of the Double-Crested Cormorant," 2006.

CHAPTER 10

1. U.S. Code Collection," 1972.
2. Price, "Hats off to Audubon," 2004.

3. Beetham, "Migratory Bird Treaty Act Turns 90," 2008.
4. Weidensaul, *Of a Feather,* 2007, 151.
5. Zimmerman, "Bald Eagle Eggs for Ten Bucks," 2008.
6. Stewart, Drew, and Wexler, "How Conservation Grew from a Whisper to a Roar," 1999.
7. "The Ramsar Convention."
8. "What Is CITES?"

CHAPTER 11

1. "Wildlife Damage Management."
2. www.aphis.usda.gov.
3. "Wildlife Damage Management."
4. "FWS Fundamentals."
5. "National Environmental Policy Act (NEPA)."
6. These states are Alabama, Arkansas, Florida, Georgia, Illinois, Indiana, Iowa, Kansas, Kentucky, Louisiana, Michigan, Minnesota, Mississippi, Missouri, New York, North Carolina, Ohio, Oklahoma, South Carolina, Tennessee, Texas, Vermont, West Virginia, and Wisconsin.
7. "Lindberg," 2007.
8. These states are Alabama, Arkansas, Florida, Georgia, Kentucky, Louisiana, Minnesota, Mississippi, North Carolina, Oklahoma, South Carolina, Tennessee, and Texas.
9. "Wildlife Damage Management."
10. "WS Directive," 2003.
11. Reed et al., "Review of the Double-Crested Cormorant Management Plan, 2003," 2003.
12. "Migratory Bird Permits," 2003.
13. Reed et al., "Review of the Double-Crested Cormorant Management Plan, 2003," 2003.
14. "Migratory Bird Permits," 2003.
15. Reed et al., "Review of the Double-Crested Cormorant Management Plan, 2003," 2003.
16. Fink, "Ecological Considerations in Fisheries Management," 2002.
17. "HSUS and Others Sue to Stop Unlawful and Unjustified Cormorant Killings," 2004.
18. "HSUS and Others Sue to Stop Unlawful and Unjustified Cormorant Killings," 2004.
19. Singh, "Court Allows Taking of Double-Crested Cormorant," 2005.
20. Casscles et al., "Challenges of Implementing the Double-Crested Cormorant Environmental Impact Statement," 2003.
21. "Cormorants Continue to Get a Bad Rap," 2007.

CHAPTER 12

1. "U.S. Aquaculture," 2007.

2. Calhoun, Reeder, and Bagl, "Federal Funds in the Black Belt," 2000.
3. Walker, "Serving Alabama Aquaculture."
4. Hanson, "Catfish Farming in Mississippi."
5. Taylor and Strickland, "Effects of Roost Shooting on Double-Crested Cormorant Use of Catfish Ponds," 2008.
6. Jerry Feist, personal communication, February 24, 2010.
7. Greg Whitis, personal communication, February 22, 2010.
8. Jerry Feist, USDA/WS figures, personal communication, February 24, 2010.
9. "Managing Wildlife Damage to Crops and Aquaculture," 2008.
10. Perez, *Fishing for Gold*, 2006.

CHAPTER 13

1. Smothers, "It's Fish Farmers vs. Cormorants, and the Birds Are Winning," 1989.
2. Line, "A Taste for Catfish Lands a Bird in Trouble as Farmers Seek to Arm," 1997.
3. Coblentz, "Cormorant Problem Tackled on Two Fronts . . . ," 2008.
4. Belant, Tyson, and Mastrangelo, "Effects of Lethal Control at Aquaculture Facilities on Populations of Piscivorous Birds," 2000.
5. Glahn, Tobin, and Blavkwell, "A Science-Based Initiative to Manage Double-Crested Cormorant Damage to Southern Aquaculture," 2000.
6. Biscobing, "Bird Watching Helps Arizona's Tourism Industry Soar," 2007.
7. Biscobing, "Bird Watching Helps Arizona's Tourism Industry Soar," 2007.
8. Glahn, Tobin, and Blavkwell, "A Science-Based Initiative to Manage Double-Crested Cormorant Damage to Southern Aquaculture," 2000.
9. Glahn, Tobin, and Blavkwell, "A Science-Based Initiative to Manage Double-Crested Cormorant Damage to Southern Aquaculture," 2000.
10. Duffy, "Why Is the Double-Crested Cormorant a Problem?," 1995.
11. Smothers, "It's Fish Farmers vs. Cormorants, and the Birds Are Winning," 1989.
12. Jerry Feist, personal communication, February 24, 2010.
13. Feist, "The Impact of the Double-Crested Cormorant on Aquaculture and Natural Fisheries," 2010.
14. Jerry Feist, personal communication, February 24, 2010.
15. Smothers, "It's Fish Farmers vs. Cormorants, and the Birds Are Winning," 1989.
16. *Economic Impact of the Mississippi Farm-Raised Catfish Industry at the Year 2003*, 2003.
17. Dr. Jim Steeby, personal communication, January 14, 2010.
18. Bennett, "Cormorant Damage Hits Ponds Hard," 2002.
19. "Resolution Regarding Management of Depredating Migratory Birds," 2004.
20. Dr. Jim Steeby, personal communication, February 24, 2010.
21. Bennett, "Agencies Differ on Cormorant Control," 2002.

22. Susan Boyles, personal communication, April 15, 2010.
23. Dr. Albert Manville, personal communication, March 17, 2010.
24. Glahn, Tobin, and Harrel, "Possible Effects of Catfish Exploitation on Overwinter Body Condition of Double-Crested Cormorants," 1997.
25. Glahn, Tobin, and Harrel, "Possible Effects of Catfish Exploitation on Overwinter Body Condition of Double-Crested Cormorants," 1997.
26. Glahn, Tobin, and Harrel, "Possible Effects of Catfish Exploitation on Overwinter Body Condition of Double-Crested Cormorants," 1997.
27. Glahn, Tobin, and Blavkwell, "A Science-Based Initiative to Manage Double-Crested Cormorant Damage to Southern Aquaculture," 2000.
28. Glahn, Tobin, and Blavkwell, "A Science-Based Initiative to Manage Double-Crested Cormorant Damage to Southern Aquaculture," 2000.
29. Hanson, "Impact of Double-Crested Cormorant Depredation on the U.S. Farm-Raised Catfish Industry," 2001.
30. "Catfish Production Declines under Economic Struggles," 2009.

CHAPTER 15

1. Paul Driessen's *Eco-Imperialism: Green Power, Black Death,* is cited under "DDT" on the African American Environmentalist Association website, www.aaenvironment.com.
2. "DDT," 2007.
3. Garrett, *The Coming Plague,* 1994.
4. Roberts, "A New Home for DDT," 2007.
5. Carson, *Silent Spring,* 1962, 25.
6. Jerry Feist, personal communication, February 24, 2010.
7. James Farquhar, NYSDEC, personal communication, January 27, 2010.
8. James Farquhar, NYSDEC, personal communication, January 27, 2010.
9. James Farquhar, NYSDEC, personal communication, January 27, 2010.
10. Mitchell Franz, personal communication, January 29, 2010.
11. Stephanie Boyles, personal communication, April 15, 2010.
12. Dr. Greg Butcher, personal communication, April 13, 2010.
13. Yauck, "The Incredible, Indelible Cormorant," 2009.
14. Bennett, "A Tale of Common Misery," 2003.
15. International Association for Great Lakes Research, "Cormorants Can Impact Fish Populations," 2008.
16. Mosedale, "Most-Hated Birds in the World," 2008.
17. "Audubon Minnesota Position on Cormorant Control in Minnesota," 2004.
18. CDI website, www.zoocheck.com/cormorant/.
19. Julie Woodyer, personal communication, March 17, 2010.
20. "The Persecution of Cormorants."
21. Julie Woodyer, personal communication, March 17, 2010.
22. CDI website, www.zoocheck.com.
23. Julie Woodyer, personal communication, March 17, 2010.

24. Ainslie Willock, personal communication, April 8, 2010.
25. Diane Pence, personal communication, March 16, 2010.
26. James Farquhar, NYSDEC, personal communication (follow-up), March 26, 2010.
27. Viney, "The Gothic World of the Much-Maligned Cormorant," 2009.
28. Viney, "The Gothic World of the Much-Maligned Cormorant," 2009.
29. Helmholtz Centre for Environmental Research (UFZ), "The Cormorant," 2008.
30. Viney, "The Gothic World of the Much-Maligned Cormorant," 2009.
31. Helmholtz Centre for Environmental Research (UFZ), "The Cormorant," 2008.
32. Helmholtz Centre for Environmental Research (UFZ), "The Cormorant," 2008.
33. Manzi and Comes, "Cormorant Fishing in Southwestern China," 2002.
34. Linda Wires, personal communication, March 8, 2010.
35. Wires and Cuthbert, "Historic Population of the Double-Crested Cormorant," 2006.
36. Linda Wires, personal communication, March 8, 2010.
37. Linda Wires, personal communication, March 8, 2010.
38. Stephanie Boyles, personal communication, April 15, 2010.
39. Leopold, *A Sand County Almanac,* 1949.
40. Shauna Hanisch, personal communication, March 19, 2010.
41. Louv, *Last Child in the Woods,* 2006.
42. Louv, *Last Child in the Woods,* 2006.
43. Louv, *Last Child in the Woods,* 2006.

Bibliography

Adoutte, A., et al. "The New Animal Phylogeny: Reliability and Implications." *Proceedings of the National Academy of Sciences of the United States of America,* April 25, 2000. www.pnas.org/content/97/9/4453.full (accessed March 17, 2009).

Ashmole, N. P. "The Regulation of Numbers of Tropical Oceanic Birds." *Ibis* 103 (1963): 458–73.

Ashworth, William. *The Late, Great Lakes: An Environmental History.* New York: Knopf, 1986.

"Audubon Minnesota Position on Cormorant Control in Minnesota." Wild River Audubon, 2004. www.wildriveraudubon.org/v25/v25_8%20doublecrestedcormorants.htm (accessed March 30, 2010).

"Battle of the Shipbuilders: The War of 1812." Great Canadian Lakes. www.greatcanadianlakes.com/ontario/lake_ontario/his_page2.htm (accessed September 8, 2009).

Beetham, John. "Migratory Bird Treaty Act Turns 90." Audubon Society of the District of Columbia, 2008. www.dcaudubon.org/node/7377 (accessed November 17, 2009).

Belant, Jerrold L., Laura A Tyson, and Philip A. Mastrangelo. "Effects of Lethal Control at Aquaculture Facilities on Populations of Piscivorous Birds." *Wildlife Society Bulletin* 28, no. 2 (summer 2000): 379–84.

Bennett, David. "A Tale of Common Misery: How Michigan Isn't Dealing with Its Cormorant Problem." *Delta Farm Press,* June 13, 2003. deltafarmpress.com/mag/farmoing_tale_common_misery (accessed March 29, 2010).

Bennett, David. "Agencies Differ on Cormorant Control." *Delta Farm Press,* March 8, 2002. http://dwltafarmpress.com/mag/farming_agencies_differ.control (accessed September 24, 2008).

Bennett, David. "Cormorant Damage Hits Ponds Hard." *Delta Farm Press,* March

1, 2002. http://deltafarmpress.com/mag/farming_cormorant_damage.hits (accessed January 23, 2010).

Berger, Cynthia. "Recycling for the Birds." National Wildlife Federation, April 1, 1994. www.nwf.org/news-and-magazines/national-wildlife/birds/archives/1994/recycling-for-the-birds.aspx (accessed March 7, 2010).

Biscobing, David. "Bird Watching Helps Arizona's Tourism Industry Soar." *Cronkite News Service,* April 26, 2007.

Board, R. G., and V. D. Scott. "Porosity of the Avian Eggshell." *American Zoologist* 20, no. 2 (1980): 339.

Bogue, Margarite Beattie. *Fishing the Great Lakes: An Environmental History, 1783–1933.* Madison: University of Wisconsin Press, 2000.

"Bullying the Government." Editorial. *New York Times,* August 20, 1852.

Butler, Patrick J. "Diving beyond the Limits." *News in Physiological Sciences* 16, no. 5 (October 2001): 222–27.

Calhoun, Samuel D., Richard J. Reeder, and Faqir S. Bagl. "Federal Funds in the Black Belt." *Rural America* 15, no. 1 (January 2000): 20–27.

Carson, Rachel. *Silent Spring.* Greenwich: Fawcett Publications, 1962.

Casscles, Kristina, et al. "Challenges of Implementing the Double-Crested Cormorant Environmental Impact Statement." USDA, 2003. www.aphis.usda.gov/wildlife_damage/nwrc/publications/03pubs/king031.pdf.

"Catfish Production Declines under Economic Struggles." Cornell University Cooperative Extension, August 11, 2009. www.extension.org/pages/catfish_declines_under_economic_struggles (accessed January 16, 2010).

Checkett, Michael J. "Understanding Waterfowl." *Ducks Unlimited,* March–April 2009. www.ducks.org/DU_magazine (accessed March 12, 2009).

"Chemical Demand Outstrips Output." *New York Times,* June 7, 1947, 34.

Chiappe, Luis M. *Glorified Dinosaurs: The Origin and Early Evolution of Birds.* Hoboken: John Wiley and Sons, 2007.

Coblentz, Bonnie. "Cormorant Problem Tackled on Two Fronts . . ." Mississippi State University/Landmarks, 2008. www.dafvm.msstate.edu/landmarks/08/winter/12-13.pdf (accessed August 26, 2008).

"Cormorants." *Green Nature.* greennature.com/article85.html (accessed August 29, 2008).

"Cormorants Continue to Get a Bad Rap." *ABC Birds.* American Bird Conservancy. 2007. www.abcbirds.org/newsandreports/stories/090616.html.

"Cormorants Culprits of Contamination." Great Lakes Sport Fishing Council, December 6, 1999. www.great-lakes.org/12-6-99.html (accessed December 8, 2008).

"Cormorants Have Fouled Island with Pollutants, a Report Says." *New York Times,* November 28, 1999.

Cornell Lab of Ornithology. "Double-Crested Cormorant." All about Birds. bna.birds.cornell.edu/species/441/articles/introduction (accessed August 26, 2008).

Cox, Caroline. "Pesticides and Birds: From DDT to Today's Poisons." *Journal of Pesticide Reform* 11, no. 4 (Winter 1991).

Custer, Thomas W., et al. "Organochlorine Contaminants and Reproductive Success of Double-Crested Cormorants from Green Bay, Wisconsin, USA." *Environmental Toxicology and Chemistry* 18, no. 6 (1999): 1209–17.

Darwin, Charles. *On the Origin of Species by Means of Natural Selection*. New York: New American Library, 2003.

"DDT." African American Environmentalist Association. 2007. www.aaenvironment.com/ddt.htm (accessed July 2, 2009).

"DDT: An Introduction." Duke University. www.chem.duke.edu/~jds/cruise_chem/pest/pest1.html (accessed April 11, 2009).

"DDT." United States Environmental Protection Agency. January 15, 2008. www.epa.gov/pbt/pubs/ddt.htm (accessed July 2, 2009).

"DDT Ban Takes Effect." EPA press release. United States Environmental Protection Agency. December 31, 1972. www.epa.gov/history/topics/ddt/01.htm (accessed April 11, 2009).

"DDT (dichlorodiphenyltrichloroethane)." Science Clarified. www.scienceclarified.com/Co-Di/DDT-dichlorodiphenyltrichloroethane.html (accessed July 2, 2009).

"DDT Extoxnet (dichlorodiphenyltrichloroethane)." Extoxnet: Extension Toxicology Network. Pesticide Information Project of Cooperative Extension Offices of Cornell University, Michigan State University, Oregon State University, and University of California at Davis (accessed July 2, 2009).

"DDT Pesticide Molecule." World of Molecules. www.worldofmolecules.com/pesticides/ddt.htm (accessed July 2, 2009).

"Decline and Collapse of Commercial Fishing." Clark Historical Library, Central Michigan University. clark.cmich.edu/nativeamericans/treatyrights/fishingdecline.htm (accessed September 11, 2009).

Dennis, Jerry. *The Living Great Lakes: Searching for the Heart of the Inland Seas*. Chapel Hill: Algonquin Books, 2003.

Dirksen, Sjoerd, and Theo Boudewijn. "Eggshell Thickness Measurements of Cormorant Eggs: Methods and Some Backgrounds." Wisconsin Cormorant Research Group Study. web.tiscalinet.it/sv2001/WI%20-%20CRSG/eggshell.htm (accessed July 2, 2009).

Dobson, Clive, and Gregor Gilpin Beck. *Watersheds: A Practical Handbook for Healthy Water*. Buffalo: Firefly Books, 1999.

"Double-Crested Cormorant." *Outdoor Alabama*. Alabama Department of Conservation and Natural Resources. www.outdooralabama.com/watchable_wildlife/what/birds/tropicbirds/dcc.cfm (accessed August 29, 2008).

"Double-Crested Cormorant Questions and Answers (Online)." *Audubon*. www.audubon.org/news/cormorant/qa.html (accessed June 6, 2009).

Duffy, David Cameron. "Why Is the Double-Crested Cormorant a Problem? Insights from Cormorant Ecology and Human Sociology." Abstract. *Colonial Waterbirds* 18 (1995): 25–32.

Dunn, Kyla. "Drastic Deformities." *Frontline: Fooling with Nature*. www.pbs.org/wgbh/pages/frontline/shows/nature/gallery/cormorants.html (accessed July 2, 2009).

Economic Impact of the Mississippi Farm-Raised Catfish Industry at the Year 2003. Mississippi State University Extension Service, 2003.

"Eggshell Texture and Structure." Provincial Museum of Alberta. virtualmuseum.ca/exhibitions/birds/pma/textstruc.htm (accessed July 2, 2009).

Ehrlich, Paul R., David S. Dobkin, and Darryl Wheye. "DDT and Birds." Stanford University. 1988. www.stanford.edu/group/standfordbirds/text/essays/DDT.html (accessed April 11, 2009).

Einstiipp, Manfred R., Russel D. Andrews, and David R. Jones. "The Effects of Depth on the Cardiac and Behavioral Responses of Double-Crested Cormorants (*Phalacrocorax auritus*) during Voluntary Diving." *Journal of Experimental Biology* 204 (2001): 4081–92.

Ellis, Julie C., Jose Miguel Farina, and Jon D. Witman. "Nutrient Transfer from Sea to Land: The Case of Gulls and Cormorants in the Gulf of Maine." *Journal of Animal Ecology* 75 (2006): 565–74.

Emerson, Edgar C., ed. "History of Henderson, NY, from Our County and Its People: A Descriptive Work on Jefferson County New York." Boston History Company, 1898. Available at history.rays-place.com/ny/henderson-ny.htm (accessed April 18, 2008).

Environmental Protection Agency. "National Environmental Policy Act (NEPA)." United States Environmental Protection Agency. www.epa.gov/compliance/nepa/ (accessed April 3, 2010).

Environmental Protection Agency. "National Environmental Policy Act (NEPA): Basic Information." United States Environmental Protection Agency. www.epa.gov/compliance/basics/nepa.html (accessed April 3, 2010).

Fahrenthold, David A. "Male Bass across Region Found to Be Bearing Eggs." *Washington Post,* September 6, 2006.

Feduccia, Alan. *The Age of Birds.* Cambridge: Harvard University Press, 1980.

Feduccia, Alan. *The Origin and Evolution of Birds.* 2nd ed. New Haven: Yale University Press, 1999.

Feist, Jerry. "The Impact of the Double-Crested Cormorant on Aquaculture and Natural Fisheries: An Alabama Perspective." Alabama Fisheries Association, 26th Annual Meeting, 2010.

Fink, Jason S. "Ecological Considerations in Fisheries Management: When Does It Matter?" *Fisheries* 27, no. 4 (April 2002): 10–17.

Fish and Wildlife Branch. "Review of the Status and Management of Double-Crested Cormorants in Ontario," 2006. Annual Report, Ontario Ministry of Natural Resources, 2006.

"The Fishing Cormorant: How the Chinese Angler Secures a Catch." *New York Times,* July 25, 1886, 2.

Frame, Ruth Elizabeth. "The Cormorant." *Saturday Evening Post* 233, no. 6 (1960): 63. Saturday Evening Post Society, used with permission.

Frederiksen, M., et al. "The Interplay between Culling and Density-Dependence in the Great Cormorant: A Modelling Approach." *Journal of Applied Ecology* 38, no. 3 (December 2001): 617–27.

Freethy, Ron. *How Birds Work: A Guide to Bird Biology.* Dorset, UK: Blandford Books, 1982.

Friedman, Amy, and Jillian Gilliland. "The Fisherman and the Cormorants of Udrost (a Norwegian Folktale)." *Times Herald-Record,* December 7, 2009, 48.

Fry, Michael D. "Unexpected Side Effects of Chemicals Acting as Hormone Mimics." Sidebar. *California Agriculture* 49, no. 6 (1995): 67.

"FWS Fundamentals." USFWS. www.fws.gov/info/pocketguide/fundamentals (accessed April 23, 2010).

Garrett, Laurie. *The Coming Plague: Newly Emerging Diseases in a World Out of Balance.* New York: Penguin Books, 1994.

Gaston, Anthony J., Ronald C. Ydenberg, and John Smith. "Ashmole's Halo and Population Regulation in Seabirds." *Marine Ornithology* 35 (2007): 119–26.

Glahn, J. F., M. E. Tobin, and B. F. Blavkwell, eds. "A Science-Based Initiative to Manage Double-Crested Cormorant Damage to Southern Aquaculture." UISDA/APHIS, 2000.

Glahn, J. F., Mark E. Tobin, and J. Brent Harrel. "Possible Effects of Catfish Exploitation on Overwinter Body Condition of Double-Crested Cormorants." USDA, 1997.

"Grand Jury to Investigate Killing of Cormorants." *New York Times,* September 1, 1998, B5.

"Great Lakes Fact Sheet." Environmental Protection Agency. www.epa.gov/grt lakes/factsheet.html (accessed September 14, 2009).

"Great Lakes Fish and Fishing: The Fishery." Great Lakes Information Network. www.great-lakes.net/teach/envt/fish/fish_1.html (accessed September 24, 2009).

Hanson, Terrill R. "Catfish Farming in Mississippi." *Mississippi History Now.* www.mshistory.k12.ms.us/index.php?id=217 (accessed January 17, 2020).

Hanson, Terrill R. "Impact of Double-Crested Cormorant Depredation on the U.S. Farm-Raised Catfish Industry." Department of Agricultural Economics, Mississippi State University, 2001.

Helmholtz Centre for Environmental Research (UFZ). "The Cormorant: The 'Black Plague' or an Example of Successful Species Conservation?" Press Release. Leipzig, 2008.

Henderson, Carrol L. *Birds in Flight: The Art and Science of How Birds Fly.* Minneapolis: Voyageur Press, 2008.

Hickey, Donald R. *The War of 1812: A Forgotten Conflict.* Urbana: University of Illinois Press, 1989.

Hirsh, Aaron E. "Guest Column: Fish Shares and Sharing Fish." *Opinionator: Exclusive Online Commentary from the Times,* February 3, 2009. opinionator.blogs.nytimes.com/author/aaron-e-hirsh (accessed February 14, 2009).

"History of Malaria during Wars." Malaria Site: All about Malaria. www.malariasite.com/malaria/history_wars.htm (accessed August 8, 2009).

Howard, L. "Family Phalacoracidae." Animal Diversity Web. University of Michigan Museum of Zoology, 2003. http://animaldiversity.ummz.umich.edu/site/accounts/information/Phalacrocoracidae.html (accessed March 01, 2009).

Howard, L. "Order Pelecaniformes." University of Michigan Museum of Zoology, 2003. http://animaldiversity.umich.edu/site/accounts/information/Pele caniformes.html (accessed February 28, 2009).

"HSUS and Others Sue to Stop Unlawful and Unjustified Cormorant Killings." Humane Society of the United States, February 21, 2004. www.hsus.org/wildlife/wildlife_news/hsus_and_other_sue_to_stop_unlawful_and_un justified_cormorant_killings (accessed January 15, 2008).

International Association for Great Lakes Research. "Cormorants Can Impact Fish Populations." International Association for Great Lakes Research, September 15, 2008. www.iaglr.org/jglr/34_3_506-523.php (accessed March 29, 2010).

James, Mark. "Kimmeridge Tidings." *Newsletter of the Friends of Kimmeridge and Purbeck Wildlife Reserve* 13 (March 2007). www. independent.ie.national-news (accessed January 13, 2010).

Johnson, James H., Russell D. McCullough, and James F. Farquhar. "Double-Crested Cormorant Studies at Little Galloo Island, Lake Ontario, in 2008: Diet Composition, Fish Consumption, and the Efficacy of Management Activities in Reducing Fish Predation." New York State Department of Environmental Conservation, Lake Ontario Annual Report, 2008, section 14, 1–11.

Johnson, James H., R. M. Ross, and J. F. Farquhar. "The Effects of Egg Oiling on Fish Consumption by Double-Crested Cormorants on Little Galloo Island, Lake Ontario, in 2003." New York State Department of Environmental Conservation, Lake Ontario Annual Report, 2003, section 15, 1–7.

Kehoe, Terrence, and Charles David Jacobson. "Environmental Decision Making and DDT Production at Montrose Chemical Corporation of California." Abstract. *Enterprise and Society* 4, no. 4 (December 2003): 640–75.

Klinkenborg, Verlyn. "The Dead Cormorants of Little Galloo Island." *New York Times,* August 8, 1998, A14.

Korfanty, C., W. G. Miyasaki, and J. L. Harcus. "Symposium on Double-Crested Cormorants: Population Status and Management Issues in the Midwest." USDA National Wildlife Research Symposia. Digital Commons, University of Nebraska, Lincoln, 1997.

Leopold, Aldo. *A Sand County Almanac.* Oxford: Oxford University Press, 1949.

Levy, Marc. "Pennsylvania Is Latest State to Put Cormorants on a Hit List." *Wild Singapore,* May 10, 2006. www.wildsingapore.com/news/20060506/0605162.htm (accessed May 10, 2010).

"Lindberg: Bill Will Create Fund Dedicated to Controlling Cormorants." Michigan House Democrats, June 8, 2007. http://109housedems.com (accessed January 12, 2010).

Line, Les. "A Taste for Catfish Lands a Bird in Trouble as Farmers Seek to Arm." *New York Times,* October 7, 1997, F9.

Louv, Richard. *Last Child in the Woods: Saving Our Children form Nature-Deficit Disorder.* Chapel Hill: Algonquin Books, 2006.

Lovorn, James R. "Upstroke Thrust, Drag Effects, and Stroke-Glide Cycles in Wing-Propelled Swimming by Birds." *American Zoologist* 41, no. 2 (April 2001): 154–65.

Lydecker, Ryck. "Game Fish Controversy Takes Wing: Cormorants Killed in New York to Save Bass Fish." *Boat/US,* March 2000.

"Managing Wildlife Damage to Crops and Aquaculture." USDA/WS, 2008. www.aphis.usda.gov/wildlife_damage/stte_report_pdfs/fy_2008 (accessed March 8, 2010).

Manzi, Maya, and Oliver T. Comes. "Cormorant Fishing in Southwestern China: A Traditional Fishery under Siege." *Geographical Review* 92, no. 4 (October 2002): 597.

Marina, J. N., A. S. King, and G. Settle. "An Allometric Study of Pulmonary Morphometric Parameters in Birds, with Mammalian Comparisons." Abstract. *Philosophical Transactions of the Royal Siciety of London. Series B, Biological Sciences* 326, no. 1231 (November 1989): 1–57.

Martin, Graham. "Gone Fishing." *Planet Earth Online: Environmental Research News,* October 20, 2008. planetearth.nerc.ac.uk/features/story.aspx?id=170 (accessed March 11, 2009).

Martin, Graham R., Craig R. White, and Patrick J. Butler. "Vision and the Foraging Technique of Great Cormorants, *Phalacrocorax Carbo:* Pursuit or Close Quarter Foraging." *IBIS: The International Journal of Avian Science* 150, no. 3 (March 2008): 485–94.

McLaughlin, Dorothy. "*Silent Spring* Revisited." Public Broadcasting System, *Fooling with Nature.* www.pbs.org/wgbh/pages/frontline/shows/nature/disrupt/sspring.html (accessed July 12, 2009).

"Michigan Commercial Fisheries Marketing and Product Development." Michigan Sea Grant. www.miseagrant.umich.edu/downloads/fisheries/07-701-fs-whitefish-marketing.pdf (accessed September 19, 2009).

"Migratory Bird Permits: Regulations for Double-Crested Cormorant Management." *Federal Register* 68, no. 195 (October 8, 2003): 58022–37.

"Minnesota Profile: Double-Crested Cormorant (*Phalacrocorax auritus*)." Minnesota Department of Natural Resources, 2001. www.dnr.state.mn.us/volunteer/mayjun01/cormorant.html (accessed March 3, 2010).

Mosedale, Mike. "Most-Hated Bird in the World: Sanctioned Killing of Cormorants Continues Unabated in Minnesota." *MinnPost,* July 16, 2008. www.minnpost.com/stories/2008/07/16/2581 (accessed March 30, 2010).

National Audubon Society. *Field Guide to North American Birds: Eastern Region.* New York: Knopf, 1994.

National Environment Research Council. "Planet Earth Online." PlanetEarthOnline, October 20, 2008. planetearth.nerc.ac.uk (accessed March 11, 2009).

"National Environmental Policy Act (NEPA)." Environmental Protection Agency. www.epa.gov/compliance/nepa (accessed April 3, 2010).

Neuman, J., et al. "Spatial and Temporal Variation in the Diet of Double-Crested Cormorants Breeding on the Lower Great Lakes in the Early 1990s." *Canadian Journal of Fisheries and Aquatic Sciences* 54 (1997): 1569–84.

"New DDT Report Confirms Data Supporting 1972 Ban, Finds Situation Improving." Press release. Environmental Protection Agency, August 11, 1975. www.epa.gov/history/topics/ddt/03.htm (accessed May 3, 2009).

New York State Department of Environmental Conservation. "Highlights and

Accomplishments, 2004/2005." Annual Report, Bureau of Fisheries, December 2005.

Oosthoek, Sharon. "Cormorant Debate: Which Part of the Ecosystem to Protect?" Canadian Broadcasting Corporation, May 27, 2009. www.cbc.ca/technology/story/2009/05/25/f-cormorants-environment-lake-erie-cull.html (accessed March 13, 2010).

"Organochlorines." United States Fish and Wildlife Service. www.fws.gov/Pacific/ecoservices/envircon/pim/reports/contaminantinfo/contaminants.html (accessed July 2, 2009).

Perez, Karni R. *Fishing for Gold: The Story of Alabama's Catfish Industry.* Tuscaloosa: University of Alabama Press, 2006.

"The Persecution of Cormorants." Cormorant Defenders International. www.zoocheck.com/cdiwebsite/cormorantswhy.shtml (accessed February 28, 2010).

Pinkey, Fred. "Endocrine (Hormone) Disruption in Fish." U.S. Fish and Wildlife Service, November 29, 2006. www.fws.gov/contaminants/issues/endocrinedisruptors.cfm (accessed May 3, 2010).

Pistis, Chuck, and Frank R. Lichtkoppler. "Michigan's Great Lakes Charter Fishing Industry in 2002." Survey. Sea Grant Great Lakes Network, 2003.

Polk, Amy. "Biologists, Sportsmen Outline Cormorant Control Plans." *St. Ignace News,* March 27, 2008.

Price, Jennifer. "Hats Off to Audubon." *Audubon,* December 2004. www.audubonmagazine.org/features0412/hats.html (accessed March 18, 2011).

Pritzl, Jeff, and Paul Peeters. "Cormorant Conundrum." *Wisconsin Natural Resources,* February 2008.

"The Ramsar Convention." Ramsar. www.ramar.org (accessed February 10, 2010).

Rebhahn, Peter. "Cormorant Controls Encouraged." Great Lakes Directory, 2001. www.greatlakesdirectory.org/zarticles/113corm.htm (accessed December 8, 2008).

Reed, Michael, et al. "Review of the Double-Crested Cormorant Management Plan, 2003." *American Ornithologists Union,* 2003. www.aou.org/committes/docs/conservationaddn5.pdf (accessed February 16, 2009).

"Resolution Regarding Management of Depredating Migratory Birds." *American Agricultural Association,* March 1, 2004. http://thehaa.net (accessed June 9, 2008).

Revkin, Andrew C. "In a Fishing Hamlet, No Grief for Cormorants: Near the Scene of a Slaughter, the Protected Birds Are Blamed for a Decrease in Jobs and Revenue." *New York Times,* August 9, 1998, 29.

Revkin, Andrew C. "9 Men Plead Guilty to Slaughtering Cormorants to Protect Sport Fishing." *New York Times,* April 9, 1999.

Revkin, Andrew C. "A Slaughter of Cormorants in Angler Country." *New York Times,* August 1, 1998, A1.

Ribak, G., T. Strod, D. Weihs, and Z. Arad. "Optimal Descent Angles of Shallow-Diving Cormorants." *Canadian Journal of Zoology* 85, no. 4 (April 2007): 561–73.

"The Rise of the Double-Crested Cormorant on the Great Lakes: Winning the War against Contaminants." Environment Canada, 1996. www.on.ec.gl.wildlife/factsheet/fs_cormorant-e.html (accessed January 10, 2008).

Ritter, Malcolm. "Dinosaur Find Raises Debate on Feather Evolution." *Yahoo News*, March 18, 2009. news.yahoo.com/s/ap_on_sc/sci_dinosaur_feathers (accessed March 19, 2009).

Roberts, Donald. "A New Home for DDT." *New York Times*, August 20, 2007.

Russell, Paul F. "Chapter I Malaria/DDT." U.S. Army Medical Department, Office of Medical History. http://history.amedd.army.mil/booksdoc/wwII/malariachapterI.htm (accessed August 8, 2009).

Rustem, Raymond, et al. *Double-Crested Cormorants in Michigan: A Review of History, Status, and Issues Related to Their Increased Population*. Reports, no. 2. Lansing: Michigan Department of Natural Resources, 2005.

Ryckman, D. P., et al. "Spatial and Temporal Trends in Organochlorine Contamination and Bill Deformities in Double-Crested Cormorants from the Canadian Great Lakes." *Environmental Monitoring and Assessment* 53, no. 1 (October 1998): 169–95.

S. R. "Birds of a Feather Sink Together." *Natural History*, March 2005.

Schuler, Mary. "Team Finds Key Mechanism of DDT Resistance in Malarial Mosquitoes." News Bureau, University of Illinois, June 16, 2008. news.illinois.edu/news/08/0616ddt.html (accessed August 8, 2009).

Scott, Jack Denton, and Ozzie Sweet. *The Submarine Bird*. New York: Putnam, 1980.

Scrivener, Leslie. "30,000 Cormorants Destroying Lakeside Park." *Toronto Star*, May 20, 2009.

"Section II: Uses of DDT." Online Ethics Center for Engineering and Research, July 6, 2006. www.onlineethics.org/resources/cases/carsonindex/2-ddtuse.aspx (accessed May 1, 2010).

Sharp, Eric. "Alien Invasion: A Great Lakes Dilemma." *National Wildlife*, August–September 2007.

Shermer, Michael. *Why Darwin Matters: The Case against Intelligent Design*. New York: Henry Holt, 2006.

Shipman, Pat. *Taking Wing: "Archaeopteryx" and the Evolution of Bird Flight*. New York: Simon and Schuster, 1998.

"Short Communications: Pebbles in Nests of Double-Crested Cormorants." *Willson Bulletin* 10, no. 1 (1989): 107–8.

Singh, Sabena. "Court Allows Taking of Double-Crested Cormorant." National Sea Grant Law Center. 2005. www.olemiss.edu/orgs/SGLC/National/SandBar/4.2cormorant.htm (accessed February 15, 2008).

Smothers, Ronald. "It's Fish Farmers vs. Cormorants, and the Birds Are Winning." *New York Times*, November 5, 1989, 24.

Stewart, Doug, Lisa Drew, and Mark Wexler. "How Conservation Grew from a Whisper to a Roar: History of US Conservation Movement." *National Wildlife*, December–January 1999.

"The Story of *Silent Spring*." Natural Resources Defense Council. www.nrdc.org/health/pesticides/hcarson.asp (accessed July 14, 2009).

Taylor, Jimmy, and Bronson Strickland. "Effects of Roost Shooting on Double-Crested Cormorant Use of Catfish Ponds: Preliminary Results." In *Proceedings of the 23rd Vertebrate Pest Conference,* 98–102. University of California at Davis, 2008.

Taylor, Susan B. "Stoker's DRACULA." *Explicator* 55 (1996). Excerpt available at www.questia.com/googlescholar.qst?docld=98491322 (accessed March 30, 2010).

"Theropod Dinosaurs: The 'Beast-Footed' Carnivorous Dinosaurs." University of California Museum of Paleontology. www.ucmp.berkeley.edu/diapsids/saurischia/theropoda.html (accessed February 16, 2010).

"Thiamine Deficiency Complex and Fish Mortality." United States Geological Survey, 2006. www.glsc.usgs.gov/files/factsheets/2006-1%/deficiency.pdf (accessed March 13, 2010).

Thomas, Dan. "Conservation Management Takes a Wild Turn: Who to Blame?" Great Lakes Sport Fishing Council. www.great-lakes.org/message.html (accessed January 7, 2008).

"Toxic Contaminants in the Environment: Persistent Organochlorines." Environment Canada, State of the Environment Infobase. www.ec.gc.ca/soer-ree/English/Indicators/Issues/Toxic/Tech_Sup/txsup4_e.cfm (accessed July 2, 2009).

"25 Years after DDT Ban, Bald Eagles, Osprey Numbers Soar." Press Release. Environmental Defense Fund, June 13, 1997. www.edf.org (accessed August 24, 2009).

"U.S. Aquaculture." NOAA Aquaculture Program, 2007. aquaculture2007.noaa.gov (accessed September 5, 2009).

"U.S. Code Collection: Taking, Killing, or Possessing Migratory Birds Unlawful." Cornell University Law School, 1972. www.law.cormell.edu/uscode/ usc_sce_16_00000703—000.html (accessed November 17, 2009).

US Department of the Interior. "Cormorant Research in Eastern Lake Ontario Leads to Improved Fish Stocking Methods: Other Finding Raise Concerns about Predation." Fact Sheet 98-4, U.S. Geological Survey. 1999.

Van Tuinen, M., et al. "Convergence and Divergence in the Evolution of Aquatic Birds." PubMed Central. www. pubmedcentral.nih.gov/articlerender.frgi?artid=1088747 (accessed March 12, 2009).

Viney, Michael. "The Gothic World of the Much-Maligned Cormorant." *Irish Times,* November 28, 2009.

Walker, Mitt. "Serving Alabama Aquaculture." Alabama Farmers Federation. www.alfafarmers.org/commodities/catfish.phtml (accessed December 12, 2009).

Walkom, Thomas. "Shooting Cormorants over Dead Trees Raises Suspicions about Liberal Motives . . ." *Toronto Star,* March 23, 2004.

"The War of 1812." Historical Association of South Jefferson. hasjny.tripod.com/id30.html (accessed September 8, 2009).

Watts, Bryan D., and Dana S. Bradshaw. "Population Expansion by Double-Crested Cormorants in Virginia." *Raven* 67, no. 2 (autumn 1996): 75–78.

Weidensaul, Scott. *Of a Feather: A Brief History of American Birding.* Orlando: Harcourt, 2007.

"What Is CITES?" CITES. www.cites.org/eng/disc/what.shtml (accessed November 11, 2009).

White, Craig R., et al. "Vision and Foraging in Cormorants: More Like Herons Than Hawks?" PLoS One, July 25, 2007. www.plos.org/article/info: doi%2F10.1371%2Fjournal.pone.0000639 (accessed August 29, 2008).

"Wildlife Damage Management: WS Mission and Goals." USDA/APHIS. www.aohis.usda.gov/wildlife_damage/about_mission.shtml (accessed April 23, 2010).

Wilford, John Noble. *The Riddle of the Dinosaur.* New York: Knopf, 1985.

Wilford, John Noble. "Scientists Discover a Dinosaur Clad in Flightless Feathers." *New York Times,* March 8, 2002. www.nytimes.com/2002/03/08/world/scientists-discover-a-dinosaur-clad-in-flightless-feathers.html (accessed March 19, 2009).

Wilgoren, Jodi. "A Bird That's on a Lot of Hit Lists." *New York Times,* January 18, 2002, A12.

Wilson, Rory P., and Marie-Pierre T. Wilson. "Foraging Behavior in Four Sympatric Cormorants." *Journal of Animal Ecology* 57 (1988): 943–55.

Wires, Linda R., and Francesca J. Cuthbert. "Historic Population of the Double-Crested Cormorant: Implications for Conservation and Management in the 21st Century." *Waterbirds* 29, no. 1 (2006): 9–37.

"WS Directive: Double-Crested Cormorant Damage Management." USDA, October 10, 2003. www.aphis.usda.gov/wildlife_damage/directives/2.330.pdf (accessed November 19, 2009).

Yauck, Jennifer. "The Incredible, Indelible Cormorant." *Bay View Compass,* November 24, 2009. bayviewcompass.com/archives/2464 (accessed March 30, 2010).

Zimmerman, E. A. "Bald Eagle Eggs for Ten Bucks." Our Better Nature, 2008. www.ourbetternature.org/mbta.htm (accessed November 17, 2009).

Index

AAEA. *See* African American Environmentalist Association
Africa, 179, 199–200
African American Environmentalist Association (AAEA), 199–200
Alabama
   catfish farming, roots in, 162, 171
   catfish operations, 159, 161, 163–64, 172, 178, 187, 201
   cormorant predation, 168–70, 178
   poverty in, 162
Alabama Fisheries Association, 178
Alabama Fish Farming Center, 169
alewife
   common name derivation, 98
   in cormorant diet, 10, 98, 100, 196
   effects of thiaminase on lake trout, 99–100
   as invasive species in Great Lakes, 88, 98, 196
   and lack of predators, 99
Allen, Rex, 113
American Bird Conservancy, 151
American Fur Company, 85
American Ornithologists' Union (AOU), 144, 147–48
American Veterinary Medical Association, 147
Andrews, Roy Chapman, 36
Animal and Plant Health Inspection Service (APHIS)
   certified fish farm facilities, 140
   goals, 137, 140, 176
   and implementation of EIS, 146, 149, 150, 151
   See also US Department of Agriculture; Wildlife Services
Animal Rights Foundation of Florida, 149, 201
Anopheles mosquito. *See* DDT
APHIS. *See* Animal and Plant Health Inspection Service
AQDO. *See* Aquacultural Depredation Order
Aquacultural Depredation Order, 143
   See also depredation orders
Archaeopteryx, 34–35, 37
Arkansas, 158, 159, 170, 172, 182, 187
Ashmole, Philip, 97
Ashmole's Halo, 6, 97–98, 101, 187
Ashworth, William, 71–72, 84, 87
Asia, 25, 73, 201
   See also China; Japan
Asian carp. *See* carp
Atlantic salmon
   extinction of in Great Lakes, 88, 110, 193, 194
   harvesting of, 82
   importance as predator, 89, 90, 212
   See also extinction, elements of

Audubon Society, 130–31, 193, 205, 208
    See also feather trade; Lacey Act
Aversa, Anthony, 113
avian ancestors, 29, 32, 35–37

Benedict, Craig A., 113
biodiversity, 81, 122–23, 139, 217–18
birdwatching, 58, 108, 175, 210
BirdWatch Ireland, 212
blue pike, 82
Bock, Walter, 137
    See also avian ancestors
Bogue, Margaret Beattie, 82–83, 216
boobies, 29, 32
Booth, Thurman, 182
Boyles, Stephanie, 183, 205, 215, 216
Boyle's law, 46–47, 51
Brant's cormorant, 31
British Ornithologists' Union, 97
Broom, Robert, 36
buffalo fish cultivation, 158
burbot, 82
Butcher, Dr. Greg, 205

Cahill, John, 111
California, 58, 59, 61, 87
Canadian Wildlife Service, 68
canals
    building of, 85, 101, 193
    Erie Canal, 85, 99
    and invasive species, 10, 88, 99, 196
    Welland Canal, 99
Cape Cod, Massachusetts, 120
Carolinian forests, 121
carp, 139, 157, 158, 179
Carson, Rachel, 58, 65–67, 192, 194, 201
Catfish Institute, 156, 163
Catskill Mountains, 83
CDI. *See* Cormorant Defenders International
Champlain, Samuel de, 72, 75, 79, 119
channel catfish
    conversion to cultivation of, 159
    and cormorants, 167–71, 172, 174
    cultivation of, 158–63
    description of, 159
    human consumption of, 156, 173, 180
    influence on cormorant repopulation, 155, 179–80, 181, 183–84, 187, 207
    marketing of, 163
charter boat industry, 90–92, 109, 191
    See also Henderson Harbor; Pacific salmon; smallmouth bass
Chiappe, Luis, 34
China, 25–26, 38, 72, 213
cisco, 82
CITES. *See* Convention on International Trade in Endangered Species Wild Fauna and Flora
classification, cormorants, 29–32
commercial fishing, Great Lakes
    beginnings of, 85–86
    cormorants and, 42, 95, 136, 145, 151
    decline of, 72, 88, 110, 114, 148–49, 196
    harvest numbers, 88, 193
    mechanization of, 25, 86–87
    See also tragedy of the commons
Congress of Racial Equality (CORE), 199, 200
Convention Between the United States and Great Britain (for Canada) for the Protection of Migratory Birds (MBTA)
    amended versions, 134, 139
    and Canada, 127
    and catfish growers, 174, 177, 184
    and conservation groups, 149–50
    cormorant protection under, 133, 139, 192, 193–94, 209
    enactment of, 127, 133
    provisions of, 127
    See also Draft Environmental Impact Statement; Final Environmental Impact Statement; Humane Society of the United States; National Environmental Policy Act; Weeks-McLean Act
Convention on International Trade in Endangered Species of Wild Fauna and Flora (CITES), 135, 194
Convention on Wetlands of International Importance (Ramsar Convention), 134–35, 194
Cook, Louis, 113

Cope, Edward Drinker, 36
CORE. *See* Congress of Racial Equality
Cormorant Defenders International
  (CDI), 118, 181, 209, 210
culling programs
  of cormorants
    Canadian, 117, 121, 210
    and catfish farmers, 180, 209–10
    and conflicting mandates, 139–40, 144
    cruelty issues, 118
    disease prevention, 118
    effects of, 106, 151, 175, 180, 208
    European, 212–13
    on Little Galloo Island, 203–4
    natural replenishment after, 151
    protection of vegetation, 121
    and reengineering nature, 122
    videos of, 209
  general, of predators, 105–6
Cuthbert, Francesca. *See* Wires, Linda

Darwin, Charles, 34, 35, 96
DDT
  acceptance of, 55, 57, 58, 65–66
  accumulation in food chain, 58, 59, 63
  agricultural use, 54, 57, 67, 199
  banning of, 64, 66–67, 194
  birth defects caused by, 64
  in cormorant excreta, 114–15
  and cormorants, 68, 79, 94, 175, 198
  court actions involving, 66
  creation of, 54, 193
  DDE, 59
  eggshell thinning, 63–64, 65, 79
  estrogenic properties of, 61, 63, 201
  insect resistance to, 60, 199–200
  as miracle insecticide, 54
  neural killing mechanism, 56–57, 60
  persistence in environment, 56–57
  physical properties of, 54–55
  production output of, 57–58
  recovery from, 67–68, 79, 94, 95, 98, 139
  renewed use of in Africa, 99–201
  robins as toxin indicators, 38, 39
  synergism with PCBs, 65

  use of in World War II, 55–56
  See also Carson, Rachel
deferred maturity, 97
DEIS. *See* Draft Environmental Impact Statement
depredation orders
  accountability in, 143, 144
  and Audubon Society, 205
  estimated cormorant take in, 144, 174
  and Humane Society of the United States, 149, 182, 205
  implementation problems of, 150
  lawsuit by conservation groups against, 149–50, 195
  and nonlethal measures, 174, 182–83
  private, 142–43, 195
  public, 142–43, 150, 195, 204
Detroit, 86, 116, 172
dichloro-diphenyl-trichloroethane. *See* DDT
diet studies, cormorants, 3–4, 100–101, 109, 147, 207, 208
dinosaurs
  archosauromorphs, 36
  eggs, 36
  feathers, 38
  See also Archaeopteryx; avian ancestors
Ditch, Ronald, 104, 113
DNA, 31–32
Draft Environmental Impact Statement (DEIS), 141, 144–45
  See also depredation orders
Driessen, Paul, 199
Duffy, D. C., 177
Dutcher, William, 131–32

eco-imperialism, 199
  See also Africa; DDT; malaria
eggshells, 53–54
  See also DDT
Endangered Species Act, 149, 205
Environmental Assessment (EA), 141
Environmental Defense Fund, 66, 67
Environmental Protection Agency (EPA), 66, 67, 73
  See also Ruckelshaus, William

EPA. *See* Environmental Protection Agency
Erie Canal. *See* canals
Europeans in North America
  as explorers, 74, 84
  as settlers, 13, 80, 82, 105
European Union, 212
extinction, elements of, 83

Farquhar, James, 202–3, 204, 211
feathers
  cormorants
    in buoyancy, 45–46
    in displays, 14, 93
    as eye tufts, 8
    as insulation, 45, 46
    use in flight, 11
    wettable, 9, 46
  dinofuzz, 38
  evolution of, 35, 38, 39
  in flight, 38–39
  See also *Archaeopteryx*; avian ancestors
feather trade
  brutality issues, 130
  and early Audubon movement, 130
  legislation, prohibition of, 132
  and women's fashions, 128–29, 193, 205
  See also Convention Between the United States and Great Britain (for Canada) for the Protection of Migratory Birds; Lacey Act; Weeks-McLean Act
FEIS. *See* Final Environmental Impact Statement
Feist, Jerry, 168–69, 177, 178
feminization of males, 61
  See also DDT
Final Environmental Impact Statement (FEIS)
  and commercial fisheries, 147
  conflicts of with scientific findings, 147
  and cormorant management, private, 145, 151
  general process of, 141, 149
  See also depredation orders

fish farming. *See* channel catfish, cultivation of
Fort, Stephen, 105
Fox River, 143
France, 54, 76, 211, 212
Franz, Mitchell, 3–4, 112–13, 123, 202, 203, 204
frigate birds, 29, 30, 32, 39
Fund for Animals, 149
FWS. *See* US Fish and Wildlife Service

Gamble, Diane, 104
gannets, 29, 30, 32
Garrett, Laurie, 87, 88
Germany, 35, 55, 211
Gobi Desert, 36–37
goby. *See* round goby
Grant, Ulysses S., 91, 139, 157
great cormorant, 31, 211
Great Lakes Sport Fishing Council (GLSFC), 114–15
Green Bay, Wisconsin, 64, 72, 116, 143, 206
Grinnell, George Bird, 129–30
group selection, 96
  See also Ashmole's Halo; Wynne-Edwards, Vero
gular pouch, 8, 14, 30, 31, 32, 37

habitat destruction, 135
Hanisch, Shauna, 216–17
Hardin, Garrett, 87
  See also tragedy of the commons
health advisories, Great Lakes fish, 115
Hemenway, Harriet, 130
  See also Audubon Society
hemoglobin, 48–49, 50, 51
Henderson, William, 75
Henderson Harbor
  commerce versus cormorants, 110
  cormorant culling, legal, 203–4
  founding of, 74–75
  importance of fishing to, 2, 92, 101–2, 104, 109
  as recreation destination, 78–79
  and rugged individualism, 111
  social movements, support for, 78

See also charter boat industry; Franz, Mitchell; Little Galloo Island; US Fish and Wildlife Service; War of 1812
Hesperonis. See avian ancestors
Hickey, Donald, 76
Hirsh, Aaron, 87
hormones, 52, 61–62, 63, 94
　See also DDT; eggshells
HSUS. *See* Humane Society of the United States
Hudson River, 85, 99, 196
Humane Society of the United States (HSUS)
　cormorant anticulling lawsuit, involvement in, 149–51, 182, 183, 205
　current goals, 205
Huxley, Thomas Henry, 36
hyperphagia, 184

insecticide burden. *See* DDT
intersex effects, 61
invasive species. *See* alewife; round goby; sea lamprey
Ireland, cormorants in, 211–12

Japan, 26, 55, 56, 92, 134

Kabot, John, 113
Kansas Department of Game and Fish, 157
Kennedy, John F., 66
Klinkenborg, Verlyn, 110–11
Krebs cycle, 49–50

Lacey Act, 130–31, 132, 193
　See also feather trade
Lake Champlain, 119
Lake Erie, 72, 88, 90, 99, 121, 210
Lake Huron, 123, 206
Lake Michigan, 64, 73, 86, 90, 206
　See also Green Bay, Wisconsin
Lake Ontario
　features of, 73
　fisheries (*see individual species names*)
　See also alewife; charter boat industry; Henderson Harbor; Little Galloo Island; round goby; sea lamprey; War of 1812
Lake Superior, 85, 208
lamprey. *See* sea lamprey
land ethic, 216
　See also Leopold, Aldo
Leech Lake, 208
Leopold, Aldo, 138, 216
Leopold, A. Starker, 138
Les Cheneaux Islands, 206
lethal controls, cormorants
　cruelty of, 118, 209–10
　See also depredation orders
Lindberg, Steven, 142
ling. *See* burbot
Link, Jason, 148
　See also depredation orders
Little Galloo Island
　and Ashmole's Halo, 98–101
　cormorant
　　colony on, 2–4, 92–94, 98 (*see also* nesting behavior, cormorant)
　　conflict with fishermen, 2, 116, 206
　　culls, legal, 202, 203
　　killing, illegal, 5, 17–19, 71, 195, 202
　　law enforcement involvement, 112–14
　　public reaction to, 105, 108–12
　　shooting, 5, 17–19, 71, 195, 202
　　See also Franz, Mitchell; Henderson Harbor
Little Pelican Island. *See* Leech Lake
Liver Bird, 23
Louv, Richard, 217
　See also nature-deficit disorder
Lyme disease, 106

malaria. *See* DDT
Manville, Dr. Albert, 183
market hunters. *See* feather trade
Marsh, Othniel Charles, 36, 37
Martin, Dr. Lee, 116
Massachusetts Audubon Society, 130
MBTA. *See* Convention Between the United States and Great Britain (for Canada) for the Protection of Migratory Birds

McCrea, David, 104, 113
McHugh, John, 4
Mekong River, 81
mers duces. *See* sweet seas
Mexico, 134, 194
Michigan
  charter boat fishing industry in, 89–90
  cormorant control, 142, 206–7
  cormorant extirpation in, 62–63
  and DDT spraying, 59
  See also charter boat industry; DDT; Green Bay, Wisconsin; Lake Michigan; Les Cheneaux Islands
Middle Island, 121, 210
Migratory Bird Treaty Act (MBTA)
  See Convention Between the United States and Great Britain (for Canada) for the Protection of Migratory Birds
Miller, David, 104
millinery trade. *See* feather trade
Minnesota, 143, 206, 207–9, 214
Mississippi
  "common misery" with Michigan, 26
  geology of, 161
  See also channel catfish
Mississippi River, 6, 12, 81, 161, 186, 197
Montrose Chemical Corporation, 58
  See also DDT
mutations, 31–32, 41, 60, 68
myoglobin, 49, 50, 51
myths, bird, 22–24

National Aquaculture Association, 170, 180
National Audubon Society. *See* Audubon Society
National Environmental Policy Act (NEPA), 141, 149
National Marine Fisheries Services (NMFS), 148
National Oceanic and Atmospheric Administration (NOAA), 148
Native American(s)
  as conservationists, 84, 85

  and European discoverers, 72, 107
  fishing
    commercial, 89–90
    expertise, 74, 84, 85, 89–90
    precontact, 84–85
  See also Champlain, Samuel de; Grinnell, George Bird; myths, bird
nature-deficit disorder, 217–18
neotropic cormorant, 31
NEPA. *See* National Environmental Policy Act
nesting behavior, cormorant, 5, 6–7, 12–17
neural killing mechanism, 56–57, 60
  See also DDT
Newcastle disease, 117–18
New York State Bureau of Fisheries, 109
New York State Department of Environmental Conservation (NYSDEC), 104, 112, 202–3
Niagara Falls, 82, 99
Nicolet, Jean, 72
NMFS. *See* National Marine Fisheries Service
NOAA. *See* National Oceanic and Atmospheric Administration
nonnative species. *See* alewife; round goby; sea lamprey
Northwest Passage, 72
nutrient transfer, 119
NYSDEC. *See* New York State Department of Environmental Conservation

On the Origin of Species. *See* Darwin, Charles
Oosthoek, Sharon, 121
Oregon, 59
Orinoco River, 81
Ostrum, John, 36
Overfishing
  cormorants, scapegoating, 143, 148, 150, 207
  effects of, 79, 86–87, 99, 100–101, 114
  See also commercial fishing; *and individual species names*

Pacific salmon
  economic impact of, 91, 109
  introduction of, 89, 192, 194
  predator, value as, 7, 101
  success of, 90–91, 101
Parks Canada, 121
Pasqu'ile Provincial Park, 117
pelecaniforms, 29–32, 38
pelicans, 13, 29, 30, 63, 179
  See also DDT
Pence, Diane, 4, 211
People for the Ethical Treatment of Animals (PETA), 182
Perez, Karni, 171
persistent insecticides. *See* DDT
PETA. *See* People for the Ethical Treatment of Animals
piping plovers, 120–21
pursuit diving, 30, 48, 51
  See also hemoglobin; myoglobin; respiratory system, cormorant

railroad expansion, 87–88
Ramsar Convention. *See* Convention on Wetlands of International Importance
respiratory system, cormorant, 8–9, 40, 43, 47–48
  See also Boyle's law; hemoglobin; myoglobin; pursuit diving
Revkin, Andrew C., 111
Revolutionary War, 66, 77
Roberts, Donald, 200
robins, effects of DDT on, 58–59
Roosevelt, Theodore, 131
round goby, 10, 100–101, 194, 213
Ruckelshaus, William, 66, 67, 68
Russell, Paul, 199

safe zones, cormorant, 122, 214
sauger. *See* blue pike
sea lamprey
  control of, 89
  effects on Great Lakes fisheries, 90, 99
  introduction into Great Lakes, 88, 193, 196

life cycle of, 89
self-regulating mechanisms, 96–97
Shipman, Pat, 34
Silent Spring. *See* Carson, Rachel
sister groups, evolutionary, 32
skeletal system, cormorant, 39–40, 45, 46, 54
smallmouth bass
  as cormorant prey, 10, 110, 184
  decline of fishery in Great Lakes, 109–10
  as favorite of Great Lakes anglers, 2, 3, 90, 204
  intersex phenomenon in, 61
Smothers, Ronald, 173, 177
snake birds, 29
songbirds, 58, 119, 128, 131
South Africa, 36, 66–67, 199
Soviet Union, 57, 134
sport fishing industry. *See* charter boat industry
state budgets, cormorant control, 142–43
Steeby, Dr. Jim, 179, 181
Stockholm Convention on Persistent Organic Pollutants, 67
sturgeon, 74, 81–82, 87, 110, 193
sweet seas, 72, 80

theropods, 35–36, 37, 38
  See also avian ancestors
Thomas, Dan, 114
Tommy Thompson Park, Toronto, 210
tragedy of the commons, 87, 88, 114, 206
tropic birds, 29, 32
typhus. *See* DDT

uncertainty principle in ecosystems, 192, 197
United Nations, 81, 100
USDA. *See* US Department of Agriculture
US Department of Agriculture (USDA)
  catfish industry, involvement in, 162, 164, 175, 176, 185–86
  goals of, 136, 137–38, 182–83, 184, 205

US Department of Agriculture (USDA) (*continued*)
  See also Animal and Plant Health Inspection Service; channel catfish; depredation orders; Draft Environmental Impact Statement; Final Environmental Impact Statement; Humane Society of the United States; Little Galloo Island; Wildlife Services
US Department of the Interior, 131, 137, 182, 213
US Fish and Wildlife Service
  catfish industry involvement, 168, 182–83
  conflicting mandates of, 144
  cormorant actions of
    selective management, 14, 116, 146–47, 148–51
    tracking, 12
  creation of, 139, 157
  mission of, 139, 144, 157, 182, 211
  See also Cape Cod, Massachusetts; channel catfish; depredation orders; Draft Environmental Impact Statement; Final Environmental Impact Statement; Little Galloo Island
USFWS. *See* US Fish and Wildlife Service

vegetation, cormorant destruction of, 16, 118–19
Vienna Convention on the Law of Treaties, 137

walleye fisheries. *See* Leech Lake
War of 1812, 76–78
  See also Henderson Harbor
Weeks-McLean Act, 132, 193
  See also feather trade

Welland Canal. *See* canals
whitefish
  harvests of, 74, 88, 90, 193
  popularity as food fish, 82, 85, 88–89, 163
  as predator, 82, 212
White-Stevens, Dr. Robert, 66
  See also DDT
Whitis, Greg, 169
Wildlife Services (WS)
  catfish farming, aid to, 168–69, 170, 176–77, 182–83, 215
  cormorant control, 142, 143, 144, 146
  history of, 137–38
  mandate of, 137, 138, 144, 205, 211
  See also Animal and Plant Health Inspection Service; depredation orders; Draft Environmental Impact Statement; Little Galloo Island; US Department of Agriculture; US Fish and Wildlife Service
Willock, Ainslie, 210
Wires, Linda, 122, 208, 214
Wisconsin fisheries, 142–43, 206
  See also Green Bay, Wisconsin
Woodley, Stephen, 121–22
Woodyer, Julie, 209, 210
World War I, 54, 56
World War II, 55–56, 58, 78
  See also DDT
WS. *See* Wildlife Services
Wynne-Edwards, Vero, 96–97

yellow perch
  as anglers' favorite, 90, 91, 204, 206–7, 208, 215–16
  as cormorant prey, 3, 10, 184, 206–8
  as food fish, 163, 204, 206